Venture Capital

European University Studies

Europäische Hochschulschriften

Publications Universitaires Européennes

Series V

Economics and Management

Reihe V Série V

Volks- und Betriebswirtschaft

Sciences économiques, gestion d'entreprise

Vol./Bd. 3381

PETER LANG

Frankfurt am Main · Berlin · Bern · Bruxelles · New York · Oxford · Wien

Acknowledgements

"The quest for wisdom starts with the genuine desire to learn."

The Bibel

Every thesis starts with a desire, the desire to wander an unknown and arduous path of learning towards a blurry goal in the future. This path has many crossings, some are misleading and need to be walked back, some are rocky or steep and hard to conquer, only few passages are easy to pass. On this journey other people are waiting on the roadside and support the wanderer to fulfil his quest - some are continuous supporters and encouragers and some just guide the wanderer for a while or give advice for which direction to take. Unless the final goal is clear and the wanderer has enough stamina he will reach his destination. But in the end he will realize that it was not the end of the journey that provides pleasure but what he experienced on his way and the individuals he met, the discussions he had which were most enjoyable.

I had the pleasure to go such a path and I enjoyed the company of many supporters whom I am deeply indebted to:

First, and foremost, I want to thank Prof. Dr. Bernd Rudolph from the Ludwig-Maximilians-University Munich, my doctoral advisor for accepting me as an external doctorial candidate, for his support, encouragement and guidance over the last two and a half years. He granted me a high level of freedom to dive-deep into the research topics I am most interested in and was a reliable source of advice. Despite my external position to his chair

I always felt welcome and part of the group when joining doctorial events, pleasure trips and Christmas parties. Therefore, I also want to thank all doctorial candidates whom I met during my MBR studies and at the chair for Capital Market Research of Prof. Rudolph for sharing their experience, motivating me to pursue and acting as a sounding board for my ideas.

I want thank Prof. Dietmar Harhoff, Ph.D. my thesis referee for his support.

I am grateful to Prof. Dr. Manfred Schwaiger and Prof. Dr. Anton Meyer who have been my advisors during the MBR studies.

My dissertation - especially my second and third study - critically depended on the access to proprietary and confidential data on private equity and venture capital funds, their investors and their investments. I especially wish to thank the CEO of the Center for Private Equity Research (CEPRES) Dr. Daniel Schmidt, the COO Dr. Philipp Krohmer and their colleagues Dr. Axel Buchner, Matthias Kling, Ulrich Häberle and Dr. Miroslav Adamov at CEPRES who have become friends both for their support with data, as well as fruitful discussions and advice during the last years.

I want to thank especially Nora Fenske and Prof. Dr. Helmut Küchenhoff from the Statistical Consulting Unit, Department of Statistics, LMU Munich for their reliable support and advice concerning statistical issues for my second study.

I am greatful to Prof. Dr. Christoph Kaserer from the Center for Economic and Financial Studies (CEFS) at Technical University Munich and to Thomas Meyer from the European Venture Capital Association (EVCA) for their support with my third study and many fruitful discussions. I am indebted to anonymous data providers who supplied data on European funds and their investors and I wish to thank Mei Niu from the British Venture Capital Association (BVCA) for challenging my results.

My thesis was supported financially by my current employer Bain & Company. I am grateful that Bain offered me the chance to exclusively focus on my thesis for two years and sponsored my living. Moreover, I thank the other doctorial candidates at Bain who went through the same ups and downs with me in parallel for sharing their experiences and challenging my ideas.

I want to express my deepest gratitude to my loving parents Ernst and Brigitte Höll and my dear uncle Prof. Dr. Rudolf Höll who shared both my times of doubt and my times of enthusiasm. They inspired me with their ideas when I was looking for a research topic and guided me during times of disorientation. They kept faith in me and encouraged me. I was allowed to draw from their experience and advice which helped me find my way through the jungle of academia.

Last but not least, I want to thank my wife Veronika Steier who supported me mentally during the last three years doing my MBR and writing my research papers.

To all of them and any other person who supported my development during the dissertation: Thank you very much!

Table of Contents

List of Figures

List of Tables

Acronyms

AIC	Akaike Information Criterion
BCI	Business Climate Index
BCPE	Box-Cox Power Exponential
BO	Buyout
BVCA	British Venture Capital Association
CAPM	Capital Asset Pricing Model
CEPRES	Center for Private Equity Research
CPV	Credit Portfolio View
CREM	Credit Risk Evaluation Model
EVCA	European Venture Capital Association
GAMLSS	Generalized Additive Model of Location, Scale and Shape
GDP	Gross Domestic Product
GP	General Partner
IPO	Initial Public Offering
IRR	Internal Rate of Return

LP	Limited Partner
M&A	Mergers and Acquisitions
MCS	Monte Carlo Simulation
MIRR	Modified Internal Rate of Return
NASDAQ	National Association of Securities Dealers Automated Quotations
NAV	Net Asset Value
OLS	Ordinary Least Square
PE	Private Equity
PME	Public Market Equivalent
SBC	Schwarz Criterion
TVE	Thompson Venture Economics
TVPI	Total Value to Paid-In
USA	United States of America
USD	United States Dollar
VC	Venture Capital
VIF	Variance Inflation Factor

Nomenclature

Symbols of Chapter 2

η	A lead investor's share in a private equity fund
γ	Share of good funds in the market
$\gamma^*(p)$	Market clearing share of good funds for a given quality detection skill level
γ_{max}	Share of good funds in the market when placement agents or lead investors support the fund management as good as possible
γ_{min}	Share of good funds in the market when placement agents or lead investors do not support the fund management
$\hat{p}_{max}$	A monopoly certifier's skill level to correctly identify a fund management's quality
A	Financial return/surplus of a private equity fund
d	A single fund's probability to get a certificate
$d(p, \gamma)$	A single fund's probability to get a certificate at a given probability to correctly identify a fund management's quality and a given share of good funds in the market
E_I	Investors' return/surplus from a private equity fund investment
e_I	An investor's expected return from a private equity fund standardized by the extent of the fund surplus

E_{PEF}	Compensation of private equity fund management
e_{PEF}	A fund management's expected return from a private equity fund standardized by the extent of the fund surplus; equals a fund's probability to get a certificate
k	A lead investor's collusion payments standardized by the extent of the private equity fund surplus
p	Probability to correctly identify a fund management's quality
p'_{max}	Second best certifier's skill level to correctly identify a fund management's quality
$p^* = p^*(\gamma)$	Market clearing detection skill level for a given share of good private equity funds in the market
p_{max}	A certifier's skill level to correctly identify a fund management's quality
p_{min}	Investors' minimum requirement for a certifier's quality detection skill
Q	Ratio between reputation gain from investors and from private equity funds
R	Reputation of a certifier
R^{std}	A certifier's total reputation gain standardized by reputation gain from private equity funds
$R_{I,b}$	A certifier's reputation loss/threat from investors for identifying a badly performing fund
$R_{I,g}$	A certifier's reputation gain from investors for identifying a well performing fund
R_I	A certifier's total reputation gain from investors

R_I^d A certifier's total reputation gain from investors in a capital over-demand situation

R_I^s A certifier' total reputation gain from investors in a capital over-supply situation

R_{PA} A placement agent's total reputation gain

$R_{PEF,b}$ A certifier's reputation loss/threat from fund managers for not certifying a fund

$R_{PEF,g}$ A certifier's reputation gain from fund managers for certifying a fund

R_{PEF} A certifier's total reputation gain from private equity funds

R_{PEF}^d A certifier's total reputation gain from private equity fund managers in a capital over-demand situation

R_{PEF}^s A certifier' total reputation gain from private equity fund managers in a capital over-supply situation

R_{RA} A rating agency's total reputation gain

s Share of funded private equity funds that leads to market clearing

W Compensation of a certifier for his service

W_C Carried interest payments which a placement agent gets for his support in the fundraising of recommended private equity funds that get capital from investors

W_F Fixed amount which a placement agent gets for his support in the fundraising of recommended private equity funds that get capital from investors

W_{LI}^C A lead investor's compensation with collusion payments

w_{LI}^{C}	A lead investor's compensation with collusion payments standardized by the extent of the private equity fund surplus
W_{LI}^{NC}	A lead investor's compensation without collusion payments
w_{LI}^{NC}	A lead investor's compensation without collusion payments standardized by the extent of the private equity fund surplus
W_{PA}	A placement agent's compensation for his service
W_{RA}	(Fixed) compensation of a rating agency for its service
x	Share of well performing funds of recommended funds
y	Share of badly performing funds of recommended funds
z	Total number of (blindly) certified private equity funds

Symbols of Chapter 3

β	vector indicating the influence of covariates on a dependent variable
δ	probability of default within a group of investments
ϵ	normal distributed regression residual with mean 0 and constant variance
λ	Box-Cox transformation parameter
μ	centrality parameter in GAMLSS regression resp. median of BCPE distribution
ν	skewness parameter in GAMLSS regression
BCPE()	Box-Cox power exponential distribution function
E()	expected value of a variable
F()	distribution function of a return measure (IRR or MIRR)

P()	function that returns a probability
π	probability of default of an investment
σ	scale parameter in GAMLSS regression
τ	kurtosis parameter in GAMLSS regression
c	(positive and negative) cashflows
CFI	cashflows from a fund to a portfolio company
CFO	cashflows from a portfolio company to a fund
d	function of an investment's duration
G	MIRR distribution of a fund
g	MIRR distribution
h	share of committed capital of a fund invested in a year
I	duration of the investment period of a fund
i	index of an individual investment
IRR	internal rate of return of an investment
j	index of a group of investments
k	number of covariates describing an investment
l	index of an investment which did not default
m	number of covariates describing a group of investments
$MIRR$	modified internal rate of return
n	total number of investments
o	number of independent observations in GAMLSS

p	return measure (IRR or MIRR)
q	index of a group of investments a new investment belongs to
r	market return resp. return that an investor gets when investing capital in his investment alternative to a Venture Capital fund
s	index of a new investment
T	maturity of an investment or a fund
t	time index
U	total multiple of all investments over the whole lifetime of a fund
u	multiple of an investment over the whole lifetime of a fund
w	dummy variable indicating if an investment defaulted (value=1) or not (value=0)
x	vector of covariates
y	vector of a dependent variable with investment returns
z	vector of the covariates for a group of investments

Symbols of Chapter 4

c	(positive and negative) cashflows
IRR	internal rate of return of an investment
NAV	net asset values of unrealized and/or partially realized investments
T	maturity of an investment or a fund
t	time index

1 Introduction

1.1 Markets for Risky Assets

Innovation is one of the keys to success and prosperity of a modern economy. Beyond the innovation which occurs in large well established companies, mainly small start-ups and young companies developing new ideas and trying new concepts often change whole markets. For example the exploiting of the internet search potential for marketing by Google changed the whole Advertising and Marketing as well as print industry. Generally speaking, young growth companies support structural industry changes securing and creating new jobs (Kortum and Lerner, 2000; Harhoff et al., 1996) in an economy.

Success and failure of start-ups are very close to each other. In a population of young companies usually only few are really successful while a big share of them fails completely. Moreover, it usually takes several years until it turns out whether a new business idea is successful or not (Chan, 1983). This reflects the high risks any person or institution faces when investing in start-ups and new business ideas. However, taking these risks is necessary to achieve the innovations mentioned above.

To be able to invest in the start-ups, to take these risks and to handle them professionally an industry - the Private Equity (PE) and Venture Capital (VC) industry - developed (mainly) during the last decades. This industry involves specialized professional players who try to mitigate the key issues in this market: Most important players are PE and VC funds. These institutions are huge capital collectors who mediate between investors willing to take high risks to achieve extraordinary returns and start-ups with new business ideas. Beyond the capital collection and investment role Private Equity funds and Venture Capital funds usually take an active role in the companies they finance to increase the value of their investments (Gormann and Salmann, 1989).

In addition to the PE funds and VC funds other players like funds-of-funds, gate-keepers, sponsors or placement agents evolved in the industry who support the capital origination process in the interface between investors and funds. Investment banks and consulting companies facilitate the investment process between funds and start-ups (also called portfolio companies once the fund invested into them).

The key issues all of these players face are information asymmetry, uncertainty and the lack of transparency. These issues lead to serious agency problems which have to be solved (at least partially) for the market to operate smoothly. The following chapters aim to shed more light on the issues, offer solutions and enhance the general level of transparency on the market, its structures and the roles of its players.

1.2 Structure of This Work

This thesis is structured into three separate studies. Following the introduction, the study in Chapter 2 investigates a situation where the **information asymmetry** issue becomes apparent. I analyze the hurdles new PE funds face when they are trying to collect capital from investors for their first fund (during fundraising) in a theoretical study:

Generally, reputation and trust are essential to manage the significant information asymmetry between fund managers and investors. A reputation has to be built over time. Thus, new fund management teams face a major market entry barrier due to the lack of trust when raising their first fund. To overcome the missing reputation, these teams may use intermediaries like rating agencies, placement agents and lead investors to certify their quality and send trust signals to investors.

I present a theoretical model of the relationship between investors, intermediaries and first-time funds. By contrasting the fundamentally different incentives of various intermediary types I develop recommendations for managers of first-time funds on which kind of intermediary to use. Two main aspects differentiate the analysis from other investigations of intermediaries' roles: First, the impact of imperfect quality detection skills of intermediaries are modeled. Second, the analysis highlights the dependency of equilibria conditions on the capital supply in the market.

I show that rating agencies may not be able to compete in the market for first-time private equity funds because they have skills which are inferior to both placement agents' and lead investors' skills. Placement agents and lead investors can influence the funds' success *and* signal the funds' quality, while rating agencies just signal funds' quality. The superior skills allow placement agents and lead investors to efficiently operate in market environments where rating agencies' skills significantly limit their possibilities.

Maintaining their own reputation motivates placement agents to balance the interests of both investors and first-time fund managers. Thus, placement agents are more suitable certification partners for first-time funds than lead investors whose reputation is less critical to their operation:

To stay in the market, placement agents may not tarnish their reputation. Beyond considering minimum requirements from investors they will certify as many funds as possible to clear the market - which is in the fund managers' best interest. This balance between investors' and fund managers' interests triggered by reputation concerns is supported by the placement agents' compensation system. Carried interest payments are critical in this respect: In a market environment with low capital supply carried interest payments motivate placement agents to support first-time funds as well as possible and clear the market. On the contrary, if a lot of capital is in the market, high carried interest payments give placement agents the incentive to rigorously select and certify fewer funds than necessary for market clearing. The reason for this is carried interest payments outweigh the investors' interests in the placement agents' objective function.

Lead investors are just interested in maximizing their own and therefore investors' return; they disregard first-time fund managers' interests who want to set up their first fund and enter the market with a high probability of success. This is not a problem in a market with a capital shortage, though in a market with plenty of capital, fund managers can offer incentives to the lead investors at other investors' expense. This behavior may motivate the lead investors to act in the fund managers' interests. A change in the lead investors' behavior is just possible when incentives reduce other investors' return below their next best alternative return. If this happens, no other investor spends capital on funds recommended by these lead investors.

In an environment with plenty of capital available and placement agents operating in the market, investors shouldn't invest in funds recommended by lead investors. Investors can expect that first-time fund managers have offered incentives to the certifying lead investors, because rational fund managers mandate placement agents in this situation. However, in an environment of capital shortage, lead investors are desirable partners for first-time funds as the lead investors may contribute a major share of the funds' capital and act in the fund managers' interest impartially and without incentives or possible collusion.

The market environment, reputation concerns, and compensation structure influence certifiers' behavior. To successfully invest in first-time funds by interpreting certifiers' signals, investors have to consider the different determinants in parallel and act accordingly.

With the consecutive study in Chapter 3 I try to solve a part of the problem of **uncertainty** in the planning and investment process of both investors and fund managers:

VC investment returns drastically vary over time as they are affected by the macroeconomic environment. The performance distribution pattern of VC direct investments is highly skewed and kurtotic. On the one hand, about one third of investments needs to be written off completely and thus generates huge financial loss. These are (at least partially) offset by a small share of "super" performing investments with more than 100% Internal Rate of Return (IRR). In this context, Chapter 3 envisages to evaluate drivers of

the performance distribution pattern of VC direct investments and presents an approach how to predict the distribution pattern for future investments. Therefore a new methodology is suggested that consists of two steps: First, I predict the probability that an investment is written off with two different methods in analogy to the credit default literature. The first method is a statistical model that is based on the *Credit Portfolio View* approach, the second method is based on the *Credit Risk Evaluation Model.* In a second step, I predict the probability distribution of returns of non-defaulting investments applying a *Generalized Additive Model of Location, Scale and Shape (GAMLSS).* The GAMLSS allows to assess influencing factors on the centrality and the dispersion parameter of the investment return distribution in parallel. Centrality and dispersion modeling is required to enable reliable predictions of return distribution percentiles.
After generally presenting the methodology, I apply it to a dataset from the *Center for Private Equity Research (CEPRES)* to test it. The predictive power in the investigated period between 1985 and 2003 is very handsome and significantly superior to an approach suggested by Buchner et al. (2010) which is closest to mine concerning its objective.

As an extension to the new prediction methodology for direct investment returns, I suggest two methodologies which allow for predicting the return distribution pattern of a portfolio of investments (e.g., in a VC fund). The first is an extension to the approach developed by Buchner et al. (2010). The second methodology which uses the Modified Internal Rate of Return (MIRR) return measurement concept is new and promises more precise and numerically stable results. It consists of the prediction of the return distribution pattern of the direct investments in each year of the fund's investment period and the prediction of the investment duration for each year's investments. The duration prediction is conducted with a statistical model that uses GAMLSS. Knowing all of a fund's investments' expected return distributions and the respective duration distribution patterns one can calculate the expected fund MIRR return distribution pattern with a Monte Carlo simulation.

Finally, the third study in Chapter 4 aims at increasing the general **transparency** on the European Venture Capital market's historic development before (1989-1998) and after (2001-2009) the Internet bubble. I investigate cashflow data on fund level provided by European investors and portfolio company data from *CEPRES*. The statistical analysis on net and gross returns shows positive returns for the years until 1998, overall negative returns for the years since 1998 but positive returns for investments in the years 2003/04. According to my data, European early stage investment performance was statistically significant below US counterparts while European later stage investments had an equal performance with the US. Within Europe, German investments seem to have been better than the other countries. The Industrials sector had the most stable positive development while the deals in Telecommunication realized the highest mean returns.

The performance differences between different funds of the vintage years after 1998 are quite low. Also the investment success of different groups of fund investors like banks, government agencies or individuals hardly differs, i.e., all investors suffered more or less equally. One exception are corporations: This group of investors realized the lowest returns of all with 4-5% lower annual returns than the rest.

I search for reasons for the negative track record in and after the bubble and find that the market seemed to have offered some attractive investment and exit opportunities – mainly in the growth cycle between 2002 and 2008. However, these opportunities were hardly exploited. The main problem in the market after 1998 was extensive capital oversupply which must have seriously diluted the quality of fund managers and financed portfolio companies. Moreover, both investors in funds as well as fund managers invested procyclically with stock markets and IPO markets which made them participate in economic downturns but hardly profit from economic recoveries.

I conclude with the finding that in order to resurrect the European Venture Capital market a healthy downsizing is required. The amount of committed capital has to be balanced with the demand for capital. Most importantly, the market has to return to a focus on quality instead of quantity both on the fund manager as well as on the portfolio company level.

2 Study 1: External Certification in the Fundraising of First-Time Private Equity Funds

2.1 Introduction

The private equity (PE) industry has mastered ways to overcome huge information asymmetries originating in the market structure. The relationship between investors seeking PE investment opportunities, private equity funds (PEFs) and portfolio companies can be characterized as a two-step principal-agent relationship: In the first step, investors (the principals) provide capital to a PEF (the agent) while it is largely unclear whether the fund managers have the right skills and ambition level to successfully invest the capital. In the second step, the fund manager (now acting as a principal) cannot be sure about the quality of the portfolio companies he invests in and the quality of the portfolio company management (the agents).

There are three main approaches the PE industry has developed to overcome information asymmetries. First, contracts between investors, fund managers and portfolio company management are designed to align their incentives and motivate them to work on the same goals, i.e., to realize a high return for the principal investors. Second, these contracts contain clauses that allow the respective principals to control and penalize their agents if they do not act as intended. Third, reputation plays a major role in decisions on every level.

A PEF with a strong reputation has a big advantage when trying to find investors for their follow-on funds. On the contrary, a major barrier to market entry for first-time PEFs or new PEF management teams is simply their lack of reputation. These players need to find ways to develop a good reputation in the marketplace. They can either rely on the reputation of the fund's parent company (if there is any), use the reputation of individual fund management team members, engage in signaling activities or get certified by external parties. These external parties can be placement agents, lead investors or other institutions that are qualified to check the PE fund management team and recommend teams with high potential to investors.

Certification of new funds only works when the certifying institutions have a good reputation and put it at stake when recommending a fund. If nothing were at stake, nobody would believe the certification signals. A problem is triggered by the fact that the certifying institutions do not want to risk lowering their reputation without expecting something in return. Thus, the compensation has to balance the institutions' risks. This reduces the reliability of the quality signal. The interesting question is whether or not and how it is possible to develop an equilibrium in which all parties are well off at the same time: The certifying institution is nominated by PEFs and is compensated for its service and its risks, investors can rely on the certification and invest in the first-time funds and the PE fund management is able to enter the market by raising enough funds to start business. Additionally it is interesting to assess which type of institution is best in signaling a new PE fund's quality.

This study investigates these questions in depth. Therefore, Section 2.2 explains general approaches to mitigate problems involved in the fundraising of (first-time) private equity funds. After motivating the idea to mandate certifying institutions, Section 2.3 elaborates on the abilities and skills of the three types of certifiers. Moreover, the agency problems involved in mandating certifiers are explained. Using a theoretical model in Section 2.4,

I compare the three types of certifiers in detail. In my model, I especially consider different market environments that entail the behavior of the intermediaries and induce different equilibria. Section 2.5 summarizes the results of the theoretical model and derives concrete action recommendations for first-time funds.

2.2 Problems and Mitigation Options in Private Equity Fundraising

The private equity market consists of two sub-markets, the venture capital market and the (leveraged) buyout market. Both sub-markets exhibit high risks for investors who expect a high risk-adjusted return (Grabenwarter and Weidig, 2005). Direct investments of investors into companies looking for equity are generally possible, however this requires either a great readiness to assume risk or special skills to be able to reduce it. The private equity industry evolved to a multi-layer expert market (Grabenwarter and Weidig, 2005) because even institutional investors have difficulties judging the exact risk of a company themselves. This is why experts called private equity funds (PEF) developed. These are intermediaries between investors and companies searching for sources of capital (in this context called *portfolio companies*). By building a diversified fund portfolio of usually 10-30 investments (Grabenwarter and Weidig, 2005) with the capital of many investors, PEF reduce the risk for each individual investor significantly (Gompers and Lerner, 2004).

Barry (1994) characterizes the relationship between investors and private equity funds as a multiple-principal-agent relationship. This relationship is on top of the principal-agent relationship between private equity fund managers and portfolio companies they invest in. Agency problems arise at the interaction between the PEF manager and both investors and portfolio companies. These problems have their origin in the high degree of uncertainty and information asymmetries inherent in the *private* equity industry where usually little information is publicly shared.

In order to mitigate agency problems, the limited partnership organizational structure has become the vehicle of choice for the majority of private equity investments. The general partner (GP) in the limited partnership, i.e., a private equity fund management, has the requisite time, experience and skills to conduct due-diligence processes to select portfolio companies and build, support and monitor them. GPs can build a high level of expertise by realizing learning effects and economies of scale. Investors join the limited partnership as limited partners (LPs), providing capital and monitoring the GPs' activities.

Gompers and Lerner (2004) characterizes the life of a private equity fund as a cycle with a fundraising phase, an investment and a divestment phase. General partners periodically - usually every 2-3 years - raise capital from investors in a closed fund with an *ex ante* predefined life-time of usually 10 years (Cook, 1998; Sahlman, 1990). The committed capital is invested in a set of portfolio companies usually within the first six years (Ljungqvist and Richardson, 2003) and before reaching the end of the fund's lifetime all investments have to be sold or written off. Knowledge, experience, skills and a nose for value allow good GPs to generate high risk-adjusted returns for their investors.

The first and main hurdle for a PEF management team is to raise enough capital to be able to start or continue its business (Prowse, 1998). The success of fundraising is determined by a diverse set of factors. Investors have to have enough capital, they must be willing to invest it in the private equity market and they have to be persuaded by a fund management team to choose their fund and not a competitor's fund.[1]

The availability of capital and the willingness to spend it on private equity investments are - among others - driven by the macroeconomy. Economic growth as well as the general stock and bond market situations impact both

1 For a detailed study on the criteria which LPs use to select a VC fund see Groh and Liechtenstein (2009).

the supply and the demand for private equity (Gompers et al., 1998; Balboa and Martí, 2007).

Investors' willingness to spend capital on a specific fund is mainly driven by the fund management's abilities to differentiate from competitors and to be persuasive. Investors face considerable adverse selection problems when they have to choose a fund. The main reason is that differences between PEFs are significant; the discrepancy between a 'median' and a 'top quartile' performing fund has historically averaged between 7 to 10 percent IRR (internal rate of return) annually (for European Funds see Kaserer and Diller, 2004). This makes a huge difference over the 10-year lifetime of a fund. The negative consequences of adverse selection are enforced because liquidating a PE fund investment prematurely is very expensive and unattractive (Lerner and Schoar, 2004). Investors have barely any options when the performance of a fund turns out badly. So the ability to identify the best funds is the key success factor in private equity investing (Harris, 2003).

2.2.1 Screening

A high quality screening process may mitigate selection problems. Investors usually start their screening activities by reading funds' private placement memorandums (PPM). A PPM summarizes all relevant information on the fund concept, the management team, organization structure, contractual terms and legal, accounting and tax structures of a fund (Sorrentino, 2003). If a fund seems appealing and matches investors' investment profile, personal meetings are scheduled to discuss more details.

As Kaplan and Schoar (2005) found out, persistence exists in returns across a sequence of funds managed by the same GP. This is why GP-LP discussions involve an extensive review of the GPs' experience and track record. Investors try to identify the links between GP activities and development of the portfolio companies of previous funds, filtering the influence of fund managers that are not with the GP team any more. Therefore LPs want to see financial statements of previous partnerships, often including cash flow history. Details like board seats held in portfolio companies by partnerships may be requested. In addition, investors study the GPs' divestment behavior in former funds, which exit channels were used with what success, and if fund managers delayed the exit of write-off investments to hide their bad performance. From all this information, investors assess rates of return for the GPs' previous trades and try to determine whether mainly skill or luck lead to the success of a fund.

Investors also study the fund concept in depth. As private equity investments require a lot of time for follow-up and value generation in portfolio companies, a GP team may only manage a limited number of investments (Kanniainen and Keuschnigg, 2003; Cumming, 2006). In consequence a huge fund volume may lead to higher valuations of initial investments and decrease expected fund returns for investors. So the relation between the number of GPs and the intended number of investments is an indicator for a fund concept being promising. Part of the concept is also the plan for regional and industry specific investments. Investors intensively study the PEF management's experience in the relevant regions and industries to assess whether the fund management team can generate an attractive deal flow and whether it knows how to judge and value companies of the fund's focus industry.

2.2.2 Self-selection and signaling

The fund governance structure and proposed terms of the partnership agreement are examined as well. Investors are looking for PEFs that agree to a significant amount of their return being tied to the performance of the fund instead of demanding a fixed salary (Sahlman, 1990). Additionally they ask PEFs to invest in concert (Schmidt and Wahrenburg, 2004) and acknowledge extensive monitoring rights for investors. The rational is that a good PEF manager should be more willing to accept restrictive conditions than a bad manager,[2] though, later investigations of private placement

2 This argument follows the signaling literature, e.g., Stiglitz (1975), Akerlof (1970), Spence (1973) or Ross (1977).

memorandums reveal that although the quality of GPs differs significantly, the contracts they write are very similar (Schmidt and Wahrenburg, 2004). According to Gompers and Lerner (1999b) this is because PEFs want to avoid signaling their quality. It seems that self-selection does not work well with private equity funds.

The fund raising and fund selection processes are very time consuming. It can take well over a year to establish a fund (Prowse, 1998; Fenn et al., 1995) and involves high search and evaluation costs both for investors and fund management teams. Sometimes investors even lack the skills required to perform fund screening professionally or PEF managers have no experience in how to raise a fund. Especially in the case of a **first-time** fund management team the sorting problem is amplified. As one US general partner puts it

> *"Fundraising is one of the biggest barriers to entry in this business. It requires stamina and relationships [...] at the end of the day a first-time fund is very hard to raise."*

(Marco, 2002). Common signaling and screening approaches that work with follow-on funds of senior fund management teams do not suit new fund management teams. These teams usually lack a long investment history and track record that investors could analyze to judge the team's quality. First-time PEFs did not build a reputation yet which gives them less incentive to act in the investors' best interests. So they find it very hard to raise their first fund.

To mitigate the serious problems of first-time funds, Gompers and Lerner (2001) suggest four strategies, Hagenmüller (2004) proposes a fifth one: First-time funds should

1. increase the amount of capital they contribute to the fund as a positive signal of commitment to future performance. The limitation to this

approach is, that managers seldom have enough capital to increase their share.

2. identify investors not only motivated by financial returns but who want to invest in a certain technology or region for image or other reasons.
3. establish an alliance with an established institution like an investment bank or other private equity group to profit from their credibility. The downside of this alternative is that GPs need to share the management of the joint fund.
4. recruit a lead investor/sponsor who provides a significant share of the fund's capital and his name and reputation as a signal for other investors to invest.
5. use a placement agent that certifies the fund management team and supports the fund raising process.

In the following analysis I focus on the intermediaries - placement agents and lead investors among others - and assess how these players can support both first-time funds and investors in the fundraising/investment processes by decreasing information asymmetry.

In the end, every first-time fund has to decide on whether to cooperate with an intermediary and which to choose. To support their decision an assessment is required which kind of intermediary is the best choice given a certain kind of situation. This is why I comparatively investigate the different intermediary types and give an advice to first-time funds on how to behave.

No literature looks at the intermediaries like placement agents and lead investors in the PE market from a theoretical point of view yet. This is why I draw results from theoretical investigations on intermediaries in general and intermediaries in other markets in the next section to evaluate the differences in the incentive situation between the intermediary types.

2.3 Intermediaries Supporting Fundraising of First-Time Private Equity Funds

2.3.1 The Role of Intermediaries

Reasons for the activities of financial intermediaries in general can be classified in terms of their roles, functions or contributions (see Chan, 1983): a) intermediaries reduce transaction costs (like costs for collection and analysis of information), b) they support portfolio optimization and asset transformation, c) they act as information producer/signaling agent and d) they perform monitoring tasks. Intermediaries who support first-time funds in their fundraising mainly act as informed agents contributing to the resource allocation and welfare of investors, i.e., I focus on reason c). This view corresponds to early works on financial intermediation, e.g., from Leland and Pyle (1977) or Chan (1983).

Intermediaries between investors and private equity funds represent another layer in the multi-layer principal-agent setting. Other authors have investigated the role of intermediaries in other roles or other markets than private equity. Mostly their studies either focus on a single type of intermediary or analyze the optimal contract of an intermediary based on certain assumptions about risk aversion, market structures and nature of goods. For example, Peyrache and Quesada (2004) conclude that intermediaries do not exist when a seller has other ways to signal the quality of his "goods". If the seller cannot reveal information another way, intermediaries always play a role. Buyers are willing to pay for information on the good's quality if information acts like insurance against low quality. Sellers are willing to pay for information transmission to buyers if they can differentiate from low quality sellers. According to Albano and Lizzeri (2001) the existance of an information generating intermediary is market efficiency enhancing; however quality is underprovided relative to full information. A paper by Kofman and Lawarree (1993) studies collusion between an auditor and a privately-informed manager at the expense of shareholders who invest in a company run by the manager and inspected by the auditor. The authors find that the goal of complete deterrence of collusion is suboptimal because increasing punishment of the manager for his collusion increases the costs for the investor.

In the following sections, I present an overview on the intermediaries supporting fundraising of private equity funds, their roles and functions. I perform a first analysis of these intermediaries' incentive situation by drawing from the results of different analysis of intermediaries in other markets. The section afterwards investigates the fundraising intermediaries in detail in a comparative theoretical study. This investigation is totally new and helps to explain existing market structures as well as to make suggestions for optimal utilization of the different intermediary types depending on the market environment.

2.3.2 Rating Agencies

Normal rating agencies as they exist in bond, mutual fund or stock markets are not active in PE markets. Nevertheless, it is generally interesting to study how rating agencies can help in private equity fundraising.

2.3.2.1 Definition

A rating agency (RA) is an intermediary that performs detailed checks of service providers (like PEFs) to rank them and to give investors advice which PEF to choose. In return the agency gets a fixed compensation.

2.3.2.2 Incentive Situation

Economic studies concerning activities of rating agencies in stock, bond and other markets have gained attention after the Enron scandal in 2001. Authors like Gras (2003) and Rosenbaum (2004) analyze the function of rating agencies like S&P, Moody's and Fitch Ratings in international markets and argue that these institutions significantly contribute to information production in capital markets. Nevertheless one has to consider the inherent agency conflicts which rating agencies are subject to.[3]

In general, both investors and PE funds can assign a rating agency to evaluate a PE fund management team. The investor community has the problem that any investor has an incentive to wait for some other investor to hire the rating agency (free-rider problem).[4] Additionally a rating agency that is hired by an investor relies on the cooperation of the fund management team to do its work which is not self-evident. As the rated institutions are interested in getting a good rating, they have an incentive to communicate untrue or sugarcoated information and deceive rating agencies (Heinke, 2000).[5]

Because investors have free-rider-problems it must be the institutions that are rated who hire and pay the rating agency.[6] According to Smith and Walter (2001) this causes potential problems. Clients may search for the agency that promises the best rating ("rating shopping", Rosenbaum, 2004) and consequently rating agencies could have an incentive to give better ratings than objective criteria suggest. Another reason for this behavior can be that rating agencies that give bad ratings multiple times in a row start losing their clients as the perceived chance of getting a good rating declines. The problem is intensified as agencies work in an environment where performance relative to competitors is important. Heinke (2000) states that rating agencies tend to give the same rating as competing rating agencies because their correct but worse rating would frighten off more potential clients than an equal rating. Turning the problem around, one could argue that rating agencies have power over their clients. They may ask for a raise in fees in exchange for a better rating, i.e., blackmail their clients (White, 2001).

3 For a comprehensive overview on multi-level principal-agent problems see Steiner and Heinke (1996).

4 If investors want the result of a fund's quality assessment to stay private, they assign a gatekeeper. Gatekeepers are intermediaries that mainly support big institutional investors like pension funds in their investment decisions.

5 Any certifier has to face this problem. Because I focus on the comparison of different intermediary types in the theoretical analysis of Section 2.4, I do not explicitly model this problem.

6 If the **rated** party chooses the rating agency, investors can realize a superior asset allocation at zero costs (Chan, 1983).

Another conflict of interest occurs between rating agencies and investors: Rating agencies are economically thinking institutions; this implies an incentive to minimize costs for evaluating a client. The consequences might be that a) it is easier for clients to deceive agencies and b) the rating does not incorporate all information necessary for an objective judgment (Heinke, 2000).

The agency problems illustrated here are mitigated by assuming that rating agencies depend significantly on their reputation (Partnoy, 1999). Reputation is what keeps them in the market. If investors do not trust in ratings any more, rating agencies lose the basis of their business. Each time an agency sends a certification signal, it puts its reputation at stake. This is why a rating agency should have a high incentive to intensively investigate fund management teams before ranking them. They may still exploit their clients and investors but they have to make sure that the extent of the exploitation does not lead to a significant loss of trust.

The existence of rating agencies has a positive effect on the market. If institutions are destined to rely on rating agencies, they will - of their own volition - increase the quality of their work in order to act within the predefined criteria that influence their rating (Hillebrand, 2001). Institutions that don't fulfill the minimum rating criteria cannot survive. Assuming the criteria are the right ones to objectively judge an institution's risk, this induces a self-adaptation and self-selection mechanism and leads to a minimum quality standard in the market.

Rating agencies contribute to the efficiency in capital markets (White, 2001). If investors can rely on an agency's signal this saves research costs to all investors. Assuming the same assessment criteria are used for any rating, investors may additionally profit from increased comparability and transparency of investment opportunities. This can improve the quality of their investment decisions. As rating agencies perform due diligences frequently they can save costs from economies of scale.

In return for their services, rating agencies usually charge fixed fees. The fixed billing structure has the advantage that conflicts of interest are diminished as the rating agency's fees are paid even if the institution gives a bad rating.

In conclusion, I can state, that a rating agency can be an interesting intermediary for a new private equity fund manager. A PEF manager who does not know about his quality relative to competitors - which is quite probable especially for first-time fund managers without a long track record - may select a rating agency

1. to get more information on his own ability to adjust the fund management team if necessary and
2. to send a signal to investors if his quality is good.

2.3.3 Placement Agents

2.3.3.1 Definition

A placement agent (PA) is '*a company that specializes in finding institutional investors that are willing and able to invest in a private equity fund or company issuing securities*' (Guliner, 2007). Placement agents have a dual role: The private equity firms (GPs) compensate them for their efforts for raising capital but their success critically depends on the investors (LPs) that need to be willing to invest into the funds the PA recommends.

Generally one can distinguish two classes of placement agents. Before 1985 there were only five placement agents in the market (Fenn et al., 1995) - all of them *independent*. Since then the number of PAs steadily increased. Attracted by appealing fees, investment banks in particular started to build their own *captive* PAs (Mills, 2005).[7]

7 Whether independent or captive placement agents are better for GPs and LPs is not clear. Usually captive PAs can rely on a broader platform and a greater network of investment banking professionals who also interact with investors concerning other investments, e.g., on fundraising for mutual funds. However, history shows that independent placement agents were more successful in raising large amounts of capital (Guliner, 2007). One of the reasons might be that the latter have less conflict of interest than captive PAs.

2.3.3.2 Function

Placement agent services have a wide range and can add value to investor-PEF-relationships in many respects. Depending on a PEF's needs, PAs may support the fund from the concept development phase until the fund is finally closed.

The main benefit from placement agents comes from their broad experience and their ability to network with investors (Cooper, 2006). When PAs are employed prior to the launch of a PE fund, they help in the *'development of comprehensive due diligence and presentation material designed to differentiate the fund in the market and recognize the expectations and demands of the prospective investors'* (Guliner, 2007; Mills, 2005). As PAs continuously support fundraising processes they can realize economies of scale. On the one hand, PEFs can leverage this experience when defining a fund concept. Because (good) PAs have transparency on investors' investment profiles, they know which concepts sell. On the other hand, a PEF with a well prepared fund concept can approach a PA to develop a proper fund raising strategy and to identify investors whose investment profile fits the fund concept. This can significantly reduce the length of the fund raising period and the workload PEFs need to dedicate to fund raising activities.

During the actual fundraising, PEFs want to raise the intended amount in the shortest period of time. PAs support the required marketing activities by pre-screening and contacting potential investors. Moreover, they coordinate and manage roadshows and investor meetings globally and assist in the preparation of responses to investors' questions. They can support negotiations of terms with key investors and help manage legal checks by attorneys (Private Equity International, 2006). As a normal PEF performs these activities only every few years this is usually not one of its key competencies. Moreover, fundraising distracts the fund management team from regular investment activities, a distraction which can be significantly avoided by employing a placement agent.

In conclusion, placement agents' key function is the raising of funds, and the certifying role they take for (first time) funds is a necessary by-product.

2.3.3.3 Incentive Situation

Placement agents in the market for private equity funds correspond to *pure* investment advisers in the market for mutual funds or insurance policies. I use the adjective *pure* because pure advisers are *just* involved in the investment decision. A broad literature has studied the incentive situation of *general* investment advisers to derive the optimal contract for such institutions. Their results are not fully transferable as general investment advisers manage the investors' capital, in addition to giving advice. This leads to a deeper relationship and more opportunities to exploit investors. Nevertheless I want to highlight some of the authors' results to elucidate my exploration of placement agents.

Like rating agencies, investment advisers have the same two fundamental agency problems: First, an adviser may capitalize on investors and not spend the effort to acquire accurate information for improving investment decisions. Secondly, an adviser may misrepresent the information he gathers to influence the investors' investment decisions and so exploit his customer. An optimal compensation scheme must resolve both issues - motivate the adviser to work hard and to give genuine advice (Liu, 2005). A study of Golec (1992) on investment adviser compensation reveals that they are usually paid a base fee and an incentive fee. The base fee increases in the investment payoffs and the incentive fee rewards the adviser if a high investment performance is achieved relative to a predefined benchmark portfolio performance. These real-world structures are in line with early theoretical studies from Holmstrom (1979) and Grossman and Hart (1983).

Like investment advisors, placement agents get a fee of 0.5-3% (usually 2%) of the funds they support to raise (Lerner, 2005; Guliner, 2007). In case a €50 million fund is raised, this already amounts to about €1 million. Placement agents are often groups of only a few individuals who share this amount. Thus, the incentive to avoid intensive fund quality assessments and also to recommend low skilled private equity fund managers to investors is significant.

Sometimes a carried interest[8] in the limited partnership is paid in addition to the fundraising fee (Fenn et al., 1995; Guliner, 2007). This may mitigate conflicts of interest and give a placement agent an incentive to favor high quality fund managers in their advice decisions. Yet, the effect can be small as the time until carried interest is paid out is long, and more determinants drive investment performance than just the fund manager's quality.

The main mitigation factor for undesirable placement agent behavior is reputation just as it is for rating agencies. A placement agent's success in the long run mainly depends on investors who are willing to invest in the funds he proposes. This is why PAs should perform very detailed due diligences before supporting PEFs in order to make sure they do not destroy their reputation by selecting a bad performing fund. But not all PAs have this responsibility to investors - some may see themselves more as salespeople (Mills, 2005).[9]

2.3.3.4 Principals

Using a placement agent has many upsides and downsides. One disadvantage is the signal which employing a placement agent sends to investors. If an established PE fund raises capital with the support of a PA, this gives the impression that the fund management expects or actually has difficulties getting money for his fund (Fenn et al., 1995). This is why placement agents are used less for regular fundraising and more in special situations as when a) an established fund wants to find investors in a region or with a group of investors where the PEF managers lack contacts, b) a fund management wants to top-up the already raised amount of capital or c) a PEF is new in the market and requires certification from a PA to replace the missing track record and reputation (Mills, 2005).

8 "Carried interest" is the name for a share of profits after the cost of investment has been returned to investors.

9 According to a study of Private Equity International (2003), 71% of subjects interviewed state that the quality of placement agents differ significantly.

In general, placement agent assignments have been on the increase. According to Hagenmüller (2004) 45% of all funds employ a placement agent to support their fundraising activities.

2.3.4 Lead Investors

2.3.4.1 Definition

Another group of intermediaries who can potentially support new fund management teams in their fund raising efforts are lead investors (LI), also called sponsors or cornerstone investors. These are investors who financially support a private equity fund by investing a significant amount of capital - either an absolute figure or a share relative to the intended fund size or the total amount raised. A lead investor publicly reveals that he supports a fund management team and acts as a point of reference to other potential investors. As a reward for his support, lead investors get favorable investment terms and/or a share in the GP's management fees and carried interest (Fenn et al., 1995).

Lead investors are not as organized as rating agencies or placement agents. In fact, they are usually normal investors who just have a special interest in supporting a new fund to raise capital. Their interest may be multi-fold. First, the supported private equity fund management can be a spin-off of the investor, e.g., if some former investment bank employees start a new PE fund (Fenn et al., 1995). Secondly, the lead investor can be a corporate investor who sponsors a fund which intends to invest in an industry the investor is interested in. If the sponsored fund is a venture capital fund, the corporate investor might be interested in the portfolio companies the PE fund supports. Finally, the LI can be a normal investor who is attracted by the favorable investment terms and has some special skills to analyze new PE funds.

2.3.4.2 Function

Lead investors perform screening of a fund management team as any investor should do before investing and may support the fundraising directly through their contact network (Hagenmüller, 2004). Because lead investors are usually one of the first investors to look at a fund management team and because they give their name and reputation to help the PEF, the fund due diligence is expected to be comprehensive. The recommendation of a lead investor can work like a certification for a fund management team and as a quality signal to investors. Additionally, the lead investor negotiates investment conditions with the GPs which may be applied to subsequent investors as well. These activities work as a first feasibility check for the conditions offered by the fund management team and tend to improve the quality of the conditions offered to other investors. During the lifetime of a PE fund, a LI may also perform more intense monitoring activities than other investors. The big share the LI holds in the fund gives him a high incentive to support the fund management and make the whole fund a success. A LI's influence can work officially through an investor committee or informally by supporting investment decisions on a one-on-one basis with the general partners.

2.3.4.3 Incentive Situation

A lead investor position seems to be quite analogous to a syndicate leader, e.g., in syndicates for venture investing or in loan syndicates. In both situations investors invest equity in a common value generation opportunity under the lead of one institution.[10] This can be advantageous for both the lead investor and the joining investors even if joining investors are inexperienced (Tykvova, 2007). Experienced investors may increase the investment's value by contributing knowledge and expertise while sharing risks with inexperienced investors. The latter can gather know-how for future deals and realize attractive returns in the syndicate partnership.

Despite significant commonalities important differences impede the transferability of results to the PE lead investor's situation.[11] Still, lead investors

10 Wright and Lockett (2003) classify inter-firm alliances into governance structures that involve or do not involve equity transfer.

11 The tasks of lead investors and syndicate leaders differ significantly. While syndicate leaders investigate targets for joint investments and are actively involved in forming the syndicate to reduce their own risk exposure, lead investors are more like certification labels. They are usually quite passive in the

may have similar hidden agendas like syndicate leaders (Meuleman et al., 2006) and may intend to exploit other investors.

2.3.4.4 Principals

Senior PEFs should not require a lead investor as they can build on an existing track record. The fund management team already cooperated with several investors and - assuming the historical performance was good - these investors can be approached for future fundraisings. When PEFs try to acquire new investors, the LPs of former funds can act as references. In consequence, lead investors are mainly useful for first-time funds - or maybe for second-time funds in case it was impossible to build a solid reputation in the meantime. The reality is that the majority (about 56%) of first-time funds and roughly 25% of fund managers who already raised more than four funds continue to use lead investors (Hagenmüller, 2004).

2.3.5 Interim Summary

Because senior PEF managers are limited in the scope that they can manage there is always a need for first-time funds in a growing market. The attractive opportunities to get high salaries as a private equity fund manager attract highly skilled and less skilled managers alike. Good managers cannot rely on their reputation and track record in order to be successful in raising their first fund and to distinguish themselves from bad managers. To fill the reputation gap, they commonly turn to intermediaries like placement agents or lead investors. These kinds of intermediaries mitigate the adverse selection problem by acting as a certifying institution for investors.

However, the intermediaries follow their own interests in their certification decisions which may contradict investors' and fund managers' interests. To avoid this conflict, different mechanisms like reputation, incentives or

fundraising process and do not intend to reduce their risk exposure but support the GPs in finding other investors for the same risk exposure. Investors in private equity funds are also not as heavily involved in the management of the investment object as venture syndicate members are in managing their portfolio company. Furthermore, lead investors and other investors do not write direct contracts like syndicate members do, which changes the possibilities of all parties involved to regulate each others' behavior.

contractual arrangements seem feasible. The following model investigates these mechanisms. Compared to existing models on intermediaries, I explicitly model the impact of imperfect certification skills and consider the three types of real-world intermediaries comparatively. I assess differences in their incentive structure and develop suggestions as to which type is best for which market conditions.

2.4 The Model

2.4.1 Analysis of a Certifier's Behavior in a Market with Unlimited Capital Supply

2.4.1.1 Assumptions and Model Structure

I investigate a neo-institutional model world with investors and private equity funds (PEF). To simplify reality I assume that all individuals decide rationally, maximize their own utility function and do not internalize the (negative) effects of their own decisions on other economic entities.

In this world investors are seeking to invest in private equity. As they do not have the right skills to successfully invest in companies themselves, they turn to the PEFs to make investment decisions and to support the value creation process within the companies. There are many PEFs in the market offering their investment services; all of them require the same amount of capital.[12] The financial return (equals the surplus) a PEF creates through his investments indicates his quality. The share γ of PEFs is good and generates a positive return A, the rest is bad and generates $-A$ with $A > 0$.[13] PEFs get a fixed payment E_{PEF} with $0 \leq E_{PEF} < A$ for their services independent of their skills. To simplify, I assume that opportunity costs (i.e., the residual utility) of PEFs and of investors are zero respectively.[14]

12 This is just a simplification of reality and does not impact conclusions. The size of the capital amount is arbitrary.

13 Individual investments are not modeled. The return of all fund investments into many portfolio companies are investigated as if it were a single investment.

14 Changing this assumption to a value greater than zero would not change the basic results of this study.

I consider the situation of first-time funds which lack an investment history and cannot rely on their reputation when raising funds. Investors may not distinguish between PEFs of different skills. If the share of good PEFs γ were greater than 0.5, investors could blindly invest in the funds. In reality the share of good first-time PEFs is usually small,[15] so I assume

$$0 < \gamma < 0.5 \tag{2.1}$$

for the whole analysis.[16]

First-time fund management teams quite often collaborate with each other in the fund's constellation for the first time, so they do not know their own strengths and weaknesses very well yet.[17] They are all confident about their skills and want to launch their first fund; thus, they turn to a certifier of their choice. As long as the certifier's fee W is lower than the PEF's expected gains from the first and future funds, the PEF will mandate the certifier to get a recommendation it can use to attract investors.[18]

Investors do not know the PEFs' quality either; but they know that only a few of them will be successful. Because established PEFs with a long track record have a limited capacity (Martí and Balboa, 2003) investors are willing to invest in new PE fund management teams. If the teams stand the test of time, investors may build successful investment cooperations.

Due to a lack of information and skills to distinguish PEFs' qualities, investors want them to be assessed by an independent third party which can

15 An investigation by Kaserer and Diller (2004) shows that in the sample of fully liquidated funds the median Excess-IRRs of European PE funds is negative in comparison to the MCSI Europe return. This shows that even in the field of well-established private equity funds more than 50% of funds generate a negative risk-adjusted return; so one can expect an even higher share of bad funds within the group of first-time funds trying to enter the market.

16 I do not investigate what happens in the - probably very rare - situation where the *majority* of first-time funds seeking capital has a high quality. Although blind investing would work out for investors in this market environment, a certifier can still facilitate the quality of investment decisions. In this situation a certifier should find it easy to step into the market but he will also face heavier competition.

17 Self-selection processes are excluded in this analysis.

18 Because investors face a free-rider problem, PEFs have to mandate the certifier.

distinguish qualities. I generally call this institution *certifier* and distinguish between

1. rating agencies (RAs),
2. placement agents (PAs) and
3. lead investors (LIs)

as discussed in Section 2.3.

Reputation PEFs are normally designed for a time-frame of 10 years. Investors may usually not find out about the fund management's quality before the end of the fund maturity due to the J-curve structure of cashflows (Grabenwarter and Weidig, 2005). A certifier can have an incentive to work in this market for a few years realizing high fees by certifying even low quality funds and give up business before the bad quality of their advice becomes evident. For example, during the private equity hype in 1998-2001 a certifier may have decided to trade short term gains from certifying most of the endless number of new PEFs that wanted to enter the market against the long-term survival of the institution. I do not want to focus on this situation in my analysis. This is why I explicitly assume:

Certifiers make decisions on the basis that they continuously want to stay in business.

In this situation I assume a reputation game. Reputation is necessary for a certifier because the market would collapse without the long-term interest in keeping the business (= maintaining a good reputation): Certifiers have an incentive to exploit their superior knowledge about PEFs' qualities at the expense of PEFs and investors. The latter can anticipate that possibility and avoid engaging in recommended funds.[19]

19 In Section 2.4.2.4 I will investigate possibilities of how a lead investor can act as certifier without the need to rely on reputation.

Reputation is necessary for a certifier:

1. to become part of the relevant set of certification partners that is mandated by PEFs and trusted by investors and
2. to survive in the competition between certifiers.[20]

Reputation of a certifier can be interpreted as the discounted return from future certification mandates. Only certifiers with a reputation $R \geq 0$ are mandated. For the first part of the analysis I assume:

If a certifier has a good reputation investors trust his signals and invest in ***all*** *funds certified by him.*[21] *PE management teams without certificate do not receive capital to start their fund.*

For a general assessment of the function of intermediaries in the market for PE fundraising, I first consider a rating agency. The RA only provides a quality signal and gets a fixed fee for its investigation and recommendation services independent of the outcome of its recommendation. This type of certifier is the most neutral form as will be illustrated in Section 2.4.2.

A rating agency gets the amount $W = W_{RA} \geq 0$ for its services and incures a cost of zero.[22] After the successful fundraising, a certified PEF becomes a GP in the relation to the investor; investors who participate become LPs. The GP invests LPs' capital in portfolio companies and realizes a return. If this return is positive the certifier gains reputation $R_{I,g} > 0$ from investors; otherwise it loses reputation $-R_{I,b} < 0$. An average certified PEF should generate a return for investors that is at least as great as his

20 This is not necessary in a monopolistic situation.

21 This assumption equals an infinite capital supply; it is modified in the analysis of capital shortage in Section 2.4.2.

22 I do not model a connection between the accuracy of certification decisions and the costs of certification, because this complicating assumption does not change fundamental insights of the proposed model. If rising certification quality increases certification costs there will probably exist a point in which a certifier's marginal return from increasing quality is overcompensated by the rising costs. This will lead to a cost optimal quality mix for certifications. Depending on the cost structure the quality mix is more or less at the investors' disadvantage.

opportunity costs of 0, so a rational investor never rewards a certifier for good certifications more than he threatens him for bad:

$$R_{I,b} \geq R_{I,g}. \tag{2.2}$$

A certifier who certifies x good and y bad funds has to expect a reputation change of $x \cdot R_{I,g} - y \cdot R_{I,b}$. If this change is negative, the RA loses reputation and has to leave the market in the long run. As long as the reputation is positive, investors and PEFs trust the certifier.

PEFs also judge a certifier's activities because they need an accredited certificate to start their business. They observe the historical and current market situation and how many PEFs seek a certificate. They know the relative frequency a certifier chooses to issue and deny certificates. PEFs prefer a certifier that is accepted by investors and issues as many certificates as possible. The reason is that a higher probability for getting a certificate equals a higher probability for PEFs to enter the market and prove their skills.

PEFs reward any certification with $R_{PEF,g} \geq 0$ and punish any non-certification with $-R_{PEF,b} \leq 0$.[23]

Limited skills of certifiers One further assumption is essential to the model:
Certifiers cannot perfectly evaluate the quality of a PEF. They have a probability $0 < p < 1$ to correctly identify the PEF's quality, i.e., discover with the probability p that a highly skilled PEF is good and with the same probability p that a low skilled PEF is bad.[24] Correctly detected good PEFs and incorrectly detected bad PEF get a certificate. This results in the probabilities of a PEF getting a certificate as illustrated in Figure 2.1. All certifiers may correctly detect first-time funds' quality with a maximum probability

23 It is not necessary to make assumptions on how $R_{PEF,g}$ and $R_{PEF,b}$ relate to each other. This will be explained in Section 2.4.1.2.

24 Statistical considerations could suggest to differenciate between incorrectly detected good funds and incorrectly detected bad funds (also called error of first and second kind). This would extremely complicate the analysis but not change conclusions fundamentally if one assumes that higher quality detection efforts from a certifier improve both the correct detection of good and bad PEFs alike.

Figure 2.1: Model structure concerning interpretation of p. The symbols used are as explained in the text and in the nomenclature

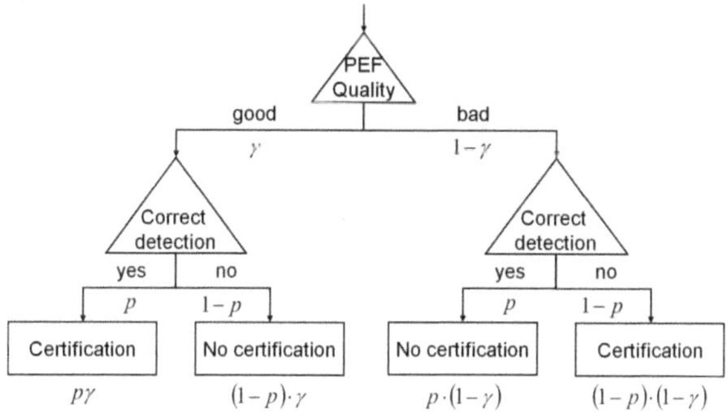

Figure 2.2: Certifier's option tree. The symbols used are as explained in the text and in the nomenclature

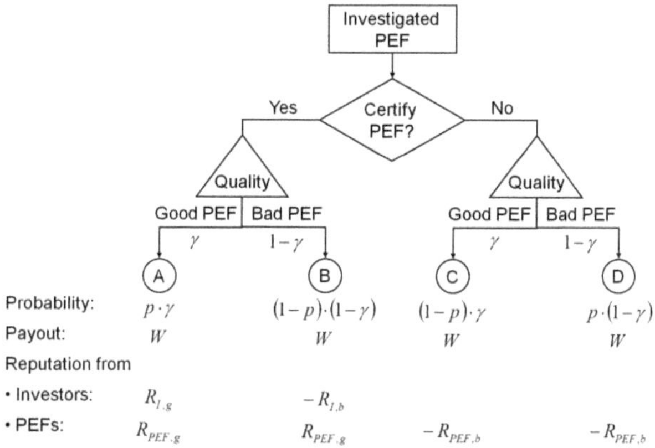

p_{max} with $0 < p_{max} < 1$. Though they may choose to investigate PEFs less accurately and realize a lower p than p_{max}.[25] Given his skills, the certifier has the option tree as illustrated in Figure 2.2.

Time structure The assumptions explained in the previous section lead to the following time structure of the model:

25 For simplicity reasons I do not model different costs for the certifier for different levels of p. I am aware that in reality this assumption cannot hold because a more precise investigation of PEFs would cause higher costs. The increasing costs with increasing p may affect the equilibrium condition motivating the PEF to realize a lower p than in the illustrated case, negatively affecting investors' return.

- $t = 0$:
  - Nature selects the quality of a new PEF, but nobody observes the quality - not even the PEF himself.
  - Certifiers offer their services to detect the PEF's quality at the price W.
  - PEFs observe the price and reputation of the certifier and decide on mandating the certifier.
  - If the certifier is mandated, he investigates the PEF's quality and sends a quality signal to investors.
- $t = 1$:
  - Investors perceive the certifier's reputation and his quality signal and decide whether they should accept the signal.
  - If investors accept it, they become LPs, invest in the fund and spend additional management costs E_{PEF} on the fund management.
  - GPs invest the capital they get from investors into portfolio companies and make every possible effort to create value in the companies.[26]
- $t = 2$:
  - Portfolio companies are sold and a return on invested capital is realized, which is dependent on the GP's quality.
  - Investors and future new PEFs observe the return and assign the certifier a reputation gain if the return is high and a reputation loss if it is low.

Please notice that extensive proofs and mathematical analysis are located in the Appendix starting from page 237 for better readability. Links to respective sections in the appendix arc in the text.

26 I explicitly do not assume a moral hazard problem on the GP's side. Considering this additional problem would only distract from the key messages of the presented model.

2.4.1.2 Equilibrium Conditions in a Polypolistic Market Environment

The average fund surplus equals the weighted sum of the value generated by good and by bad PEFs that receive capital from investors. With the probability $p \cdot \gamma$ a fund management is good and receives capital; with the probability $(1-p) \cdot (1-\gamma)$ a fund management is bad and receives capital. So the average fund surplus is $p\gamma \cdot A + (1-p)(1-\gamma) \cdot (-A)$.

I first investigate a polypolistic market where all parties push each other to their reservation utility, i.e., *many* investors want to invest capital into *many* PEFs recommended by *many* certifiers. This implies that the group with the strongest bargaining power extracts all surplus. In the current model investors are the strongest group as both first-time funds and certifiers may only exist when investors spend their capital on the certified PEFs. The whole surplus goes to investors, i.e., investors have a return E_I of

$$E_I = p\gamma \cdot A + (1-p)(1-\gamma) \cdot (-A). \tag{2.3}$$

PEFs get a management fee of $E_{PEF} = 0$ and certifiers receive a compensation of $W = 0$ from PEFs for their services.[27] The question is what are the minimum requirements for a stable signaling equilibrium in this market.

A certifier who wants to stay in business has to consider investors' and PEFs' opinions. I first investigate how the two parties steer the certifier through reputation separately and combine both views afterwards.

Requirements from investors

Ratio between certified good and bad PEFs For a certifier to remain in business the following condition has to hold:

27 In a market where capital is the bottleneck, i.e., too many funds are seeking too little capital, investors have a strong and PEFs have a weak bargaining position. Investors may always threaten a PEF with a plan to spend capital on another PEF. So a fund agrees to work for the same as its next best alternative compensation. The argumentation with certifiers is similar: A certifier's recommendations are only valuable for PEFs if investors accept them. Investors can threaten a certifier to decline his and just accept a competitor's recommendations. Thus, certifiers just charge as much as their next best alternative compensation to PEFs which recharge the same amount from investors. (For a more detailed analysis of surplus distribution in a principal-intermediary-agent setting see Lizzeri (1999).)

Lemma 1. *When a certifier wants to keep his reputation, the number of certificates for good PEFs x has to be larger than or equal to the number of certificates for bad PEFs y.*

Proof. See Appendix 1 on page 237. □

Blind certification A certifier can decide to recommend PEFs without checking their quality. In reality this potentially saves investigation costs. (In the current model, the costs are assumed to be 0 anyway.) Concerning the question of whether this is an option for the certifier, I can state:

Lemma 2. *In an environment with $0 < \gamma < 0.5$ the certifier may not blindly issue certificates if he does not want to lose reputation.*

Proof. See Appendix 2 on page 237. □

Certification after PEF investigation Blind certification is not an option, so investigating PEFs before recommending them should increase a certifier's chance to gain reputation. A certifier's quality detection skill p has a cap at p_{max}. The question is what the minimum skill level is in a market with $\gamma < 0.5$ and if the maximum skill level is sufficient for the certifier to gain reputation. To investigate these questions I consider the following example:

Example 1. *I regard a market with 100 funds seeking capital. The share of good PEFs is* 30% *and three certifiers can detect fund qualities with different probabilities. Certifier 1 has a probability of* 90%, *2 has* 60% *and 3 has* 70%. *Certifier 1 advises investors to invest in* $p \cdot \gamma \cdot 100 = 0.9 \cdot 0.3 \cdot 100 = 27$ *good funds and* $(1-p) \cdot (1-\gamma) \cdot 100 = 0.1 \cdot 0.7 \cdot 100 = 7$ *bad funds.* 66 *funds do not get a certificate. Even assuming that investors reward and punish correct and incorrect certifications equally (so* $\frac{R_{I,g}}{R_{I,b}} = 1$*) the certifier will gain reputation due to Lemma 1 although* $\gamma < 0.5$. *Certifier 2 recommends* 18 *good and* 42 *bad PEFs, in total* 60, *far more than the first certifier, but he loses reputation because the reputation loss from bad funds is not compensated by good fund certifications* ($\frac{18}{42} < 1$). *Certifier 3 operates exactly at the borderline of the* *previous cases. He certifies 21 good and 21 bad funds and does not lose reputation, if* $R_{I,g} = R_{I,b}$.

The general condition for p is:

Theorem 3. *In a stable equilibrium the probability p has to be at least greater or equal to* $p_{min} = 1 - \gamma > 0.5$*, if a certifier does not want to lose reputation.*

Proof. See Appendix 10 on page 243. □

The condition $p \geq p_{min} = 1 - \gamma$ means that a certifier's selection skills have to compensate at least for the risk $1-\gamma$ that exists in the market. If the risk increases (i.e., γ decreases), certifiers need to be able to compensate for this to stay in business. At $p = 1-\gamma$ the surplus in the market is 0 because the share of certified good and bad funds is equal, i.e., $p\gamma = (1-p)(1-\gamma)$.

Investors may choose to punish and reward a certifier unequally, which changes the minimum requirements for his certification decisions. So a certifier is interested in the relation between p, γ and $R_{I,g}$, $R_{I,b}$ when calculating the consequences of his certification actions on reputation:

Lemma 4. *If a certifier wants to keep his reputation with investors, he has to choose a level of p that holds the following conditions:*

$$p \geq \frac{1}{1 + \frac{\gamma}{1-\gamma} \cdot \frac{R_{I,g}}{R_{I,b}}} \tag{2.4}$$

Proof. See Appendix 4 on page 239. □

Ceteris paribus, investors can lift the participation condition for certifiers by disproportionally increasing the reputation threat $R_{I,b}$ over $R_{I,g}$. How effective a marginal increase is depends on the level of γ in the market (see Inequation (2.4)).

Lifting the participation condition for certifiers is an interesting option for investors. This forces certifiers to act entirely in accordance with the investors' will, though, this is only possible for a monopoly investor. In an investor polypoly there is always an investor in the market who believes a certifier who recommends funds with an average return equal to his opportunity return of 0. Thus, investors believe a certifier as long as the recommended bad funds do not outnumber the recommended good funds. As a consequence, investors' reputation reward for good funds equals the threat for bad funds, i.e., $R_{I,g} = R_{I,b}$. The certifiers' participation condition turns to $p \geq p_{min} = 1 - \gamma$.

For the following analysis I assume polypolistic market conditions for investors.

Requirements from private equity funds A certifier also has to consider the PEFs' reputation rewards $R_{PEF,g}$ and punishment $R_{PEF,b}$. I define

$$R_{PEF} = (p\gamma + (1-p)(1-\gamma)) \cdot R_{PEF,g} - (p(1-\gamma) + (1-p)\gamma) \cdot R_{PEF,b}$$

as the reputation a certifier gets from PEFs. This means, the PEF rewards a certifier with a reputation gain with the probability of getting a certificate $(p\gamma + (1-p)(1-\gamma))$ and threatens the certifier's reputation with the probability of denying certification $(p(1-\gamma) + (1-p)\gamma)$.

Theorem 5. *The PEFs' reputation change R_{PEF} is monotonic decreasing in p for the relevant interval* $0.5 < p < 1$.

Proof. See Appendix 5 on page 239. □

Theorem 5 shows that PEFs want the certifier to choose a low level of p because PEFs prefer to get a higher chance to prove their skills. With an increasing p the share of certified PEFs decreases. Depending on the power of investors vs. PEFs, the reputation game with PEFs potentially leads a certifier to choose a level of p that is as low as possible but still accepted by investors. At this point the reputation from PEFs is maximized, because a lower level of p would cause investors to reject the certifier's recommendation. A detailed analysis of this follows in the next subsection.

Simultaneous consideration of reputation from investors and private equity funds Theorem 3 and 5 frame a certifier's decision; he has to balance conditions from investors and PEFs. To analyze this, I combine both parties' reputation conditions:

1. For investors I combine the reputation aspects $R_{I,g}$ and $R_{I,b}$ to R_I. R_I equals the reputation surplus for certifying more good than bad PEFs:

$$R_I \cdot (p\gamma - (1-p)(1-\gamma)) = p\gamma \cdot R_{I,g} - (1-p)(1-\gamma) \cdot R_{I,b}$$
$$R_I = \frac{p\gamma}{p\gamma - (1-p)(1-\gamma)} \cdot R_{I,g} - \frac{(1-p)(1-\gamma)}{p\gamma - (1-p)(1-\gamma)} \cdot R_{I,b} \quad (2.5)$$

   If the investors' reputation reward and threat are equal, investors' overall reputation change $R_I = R_{I,g} = R_{I,b}$ (follows from Equation (2.5)).
2. PEFs' reputation reward for fund certification equals the negative threat for non-certification. Thus, it is sufficient to assume - without losing generalization - that PEFs threaten a certifier for non-certification, i.e., $R_{PEF,g} = 0$ and $R_{PEF} = R_{PEF,b}$.

In consequence, the choice of p and γ affects a certifier's reputation change according to the joint reputation function:

$$R = (p\gamma - (1-p)(1-\gamma)) \cdot R_I - (\gamma(1-p) + p(1-\gamma)) \cdot R_{PEF}. \quad (2.6)$$

Based on the assumptions of this model, certifiers cannot recommend more than 50% of the first-time PEFs (see Appendix 6 on page 240). Hence, if the reputation threat R_{PEF} from PEFs is greater than the reward R_I from investors, a rating agency always loses reputation. To say it differently, a reputation gain is only possible if[28]

$$\frac{R_I}{R_{PEF}} \geq \frac{p+\gamma-2p\gamma}{p+\gamma-1} > 1\,. \tag{2.7}$$

I assume that this condition is fulfilled for the next analysis; otherwise certifiers will not play in the market, and this cannot be in the PEFs' best interest.

For better illustration, I standardize the reputation change R by R_{PEF}:

$$R^{std} = \frac{R}{R_{PEF}} = (p\gamma - (1-p)(1-\gamma)) \cdot Q - (\gamma(1-p) + p(1-\gamma)). \tag{2.8}$$

with $Q = \frac{R_I}{R_{PEF}}$ equaling the ratio between reputation change from investors and PEFs.

Theorem 6. *In a market where all certified PEFs (with licenses from well respected certifiers) get capital from investors, the PEFs' reputation threat cannot deter certifiers from increasing p. Certifiers will act in investors' best interest and diligently recommend PEFs with* p_{max}*. A higher ratio* $Q = \frac{R_I}{R_{PEF}}$ *allows certifiers with a lower skill level* p_{max} *to play in the market.*

Proof. Differentiating R^{std} with respect to p returns $\frac{\partial R^{std}}{\partial p} = Q - 1 + 2\gamma$. For $\gamma < 0.5$ and $Q > 1$ this equation is always positive, so an increase in p will always yield a positive reputation to certifiers. Solving the inequation $R^{std} \geq 0$ for p returns

$$p \geq \frac{\gamma - \gamma Q + Q}{Q - 1 + 2\gamma}. \tag{2.9}$$

To determine the dependency of p from Q the partial derivative equals

$$\frac{\partial p}{\partial Q} = \frac{-2\gamma^2 + 2\gamma - 1}{(Q - 1 + 2\gamma)^2} < 0\,. \tag{2.10}$$

Thus, an increase in Q leads to a decrease in p. This result allows predictions on the relationship between certifier, PEFs and investors. An investor who disproportionally increases his reputation reward R_I enables more certifiers

28 This condition can be deduced by setting the term (2.6) for reputation greater or equal to 0.

to operate in the market as a lower p_{max} becomes sufficient. The same is true for a decrease in the reputation thread R_{PEF} from PEFs. On the contrary, if an investor decreases his reward for correct certifications or PEFs increase their threat for non-certification, the certifier has to realize a higher p. The latter case is not intuitive but directly deductible from Equation (2.10): A PEF that increases its threat makes the investors' reward for a high p less valuable, so the certifier has to increase p further in order to keep the same reputation level. □

2.4.1.3 Equilibrium Conditions in a Monopolistic Market Environment

The first part of the analysis was about a polypolistic market environment, implicitly assuming that all certifiers in the market are able to realize the same p_{max}. The question is what happens if one certifier, called a monopoly certifier (MC), can realize a higher $\hat{p}_{max} > p_{max}$. For reasons of survival he does not have to choose $\hat{p}_{max}$. With p_{max} he receives the same reputation as his colleagues as well as an income equaling his residual utility[29]. Recommending funds with $\hat{p}_{max}$ has two effects:

Lemma 7. *If the certifier uses a higher $\hat{p}_{max} > p_{max}$ the share of PEFs he can certify decreases.*

Proof. See Appendix 7 on page 240. □

In consequence to Lemma 7, PEFs dislike a certifier to increase p to $\hat{p}_{max}$ because their probability of getting a certificate decreases. This makes them devalue the certifier's reputation.

Lemma 8. *At $\hat{p}_{max} > p_{max}$ the ratio between good and bad certified PEFs increases. This leads to an increase in the market surplus[30].*

Proof. See Appendix 8 on page 241. □

29 The *residual utility* is definded as the utility of an institution's second best option.

30 *Market surplus* means the value generated by fund certification after subtracting related costs.

Lemma 7 and Lemma 8 imply that the share of certified investment opportunities in PEFs decreases with increasing p but the quality of them improves, which results in an increase in investors' return if certification costs W stay the same. This motivates investors to exclusively believe in MC's signals; PEFs are forced to mandate MC. If a second certifier can also realize $\hat{p}_{max}$, the level of p_{max} increases to $\hat{p}_{max}$ and none of the certifiers can derive an advantage from their superior skills. If MC is the only one that can certify with the probability $\hat{p}_{max}$, he becomes a monopolist. In this case, reputation is not driving his decisions any more as the reasons for reputation (see page 31) are no longer crucial. The monopolist will not compete with other certifiers as long as the return he generates for investors is higher than the return they can expect from his competitors.

Analogous to Lizzeri (1999) the monopoly certifier will try to get the full surplus generated by PEF certification, but this is not possible:

Theorem 9. *A monopoly certifier may (only) extract the surplus that is generated from the difference between p'_{max} and $\hat{p}_{max}$, where p'_{max} is the maximum skill level of the second best certifier. The monopoly certifier charges*

$$(\hat{p}_{max} - p'_{max}) \cdot A \tag{2.11}$$

from each PEF and they charge

$$(\hat{p}_{max} - p'_{max}) \cdot A \cdot \frac{1}{\hat{p}_{max}\gamma + (1 - \hat{p}_{max})(1 - \gamma)} \tag{2.12}$$

with $\frac{1}{\hat{p}_{max}\gamma+(1-\hat{p}_{max})(1-\gamma)} > 2$ *to investors.*

Proof. I start with the situation that MC realizes $\hat{p}_{max}$ without charging more than the polypol rate $W = 0$ for his service. In this case, investors just believe the MC; all PEFs require an MC-certificate to enter the market. From other certifiers - also willing to operate in the market - investors expect to get a surplus $A \cdot (p'_{max}\gamma - (1 - p'_{max})(1 - \gamma))$. Due to Lemma 8 the surplus in the market increases with $\hat{p}_{max} > p'_{max}$ to $A \cdot (\hat{p}_{max}\gamma - (1 - \hat{p}_{max})(1 - \gamma))$. The monopoly certifier creates additional value for investors, therefore he will consider charging higher fees. In a stable equilibrium he may do so because the PEF can charge higher fees from his investors if he holds an MC certificate instead of a certificate from another certifier. An MC may increase his fees further up to the point where investors are no longer performing better when spending their capital only on funds certified by him. This point is reached when the total additional surplus he creates by choosing $\hat{p}_{max}$ versus p'_{max} goes into his pocket. If he charges more to his principal than the difference in surplus between monopolist and second best certifier (equals $A \cdot (p'_{max}\gamma - (1-p'_{max})(1-\gamma)) - A \cdot (\hat{p}_{max}\gamma - (1-\hat{p}_{max})(1-\gamma)) = A \cdot (\hat{p}_{max} - p'_{max})$) his competitor would be mandated.

As illustrated, a PEF that assigns the MC pays $A \cdot (\hat{p}_{max} - p'_{max})$. He himself needs to claim something back from investors if he does not want to lose money. Because the PEF is certified only with a probability $\hat{p}_{max}\gamma + (1-\hat{p}_{max})(1-\gamma)$ he has to increase the amount to $\frac{1}{\hat{p}_{max}\gamma + (1-\hat{p}_{max})(1-\gamma)} \cdot A \cdot (\hat{p}_{max} - p'_{max})$. The maximum share a certifier may recommend is 0.5 (see Lemma 4). So PEFs charge at least twice the surplus to investors. As investors only pay fees to PEFs if the latter are certified by an MC, they are willing to pay the higher amount. □

Example 2. *To illustrate the theoretical explanations I come back to example 1 with 100 funds seeking financing and a $\gamma = 0.3$ in the market. There is a certifier Z with a maximum skill level $p_{max} = 0.8 > 1 - \gamma = 0.7$ who recommends 24 good and 14 bad PEF to investors, enabling them to realize a surplus $(24 - 14) = 10$. Another certifier $\hat{Z}$ is better and can realize a $\hat{p}_{max} = 0.9$, i.e., he recommends 34 funds, 27 good and 7 bad ones. The surplus in his recommendations is $27 - 7 = 20$, double the amount of Z. So, investors will only listen to $\hat{Z}$ and PEFs will only assign him. $\hat{Z}$ knows about the superiority of his skills and that investors earn more when listening to him than to others. His certificates are worth 10 more to investors than Z-certificates, so investors are willing to pay the difference to PEFs. $\hat{Z}$ anticipates that. He knows that all 34 certified PEFs can charge investors $\frac{10}{34}$ per fund, so the certifier thinks about charging the same to PEFs. With this behavior the certifier would receive $100 \cdot \frac{10}{34}$. Because PEFs cannot be* *sure to get a certificate, they would expect a loss of* $34 \cdot \frac{10}{34} - 100 \cdot \frac{10}{34} = -66 \cdot \frac{10}{34}$ *in this case. So the certifier may charge all PEFs only* $\frac{34}{100} \cdot \frac{10}{34} = 0.1$. *In this case he receives* 10 *in total and investors earn* $20 - 34 \cdot \frac{1}{3.4} = 10$ *which is the same as with* Z. *On average PEFs get nothing, but certified PEFs are left with a gain of* $\frac{10}{34} - \frac{10}{100} = \frac{33}{170}$. *Their gains are cross-subsidized by the non-certified funds which make a loss.*

Empirical evidence reveals that private equity funds divide into two distinctive groups, venture capital and buyout funds; both mainly invest in a few sectors and industries, e.g., hardware, software, Internet and biotech companies. Special skills and knowledge are required for each of the different industries. Focusing on a few industries should support successful fund investing.[31] The same is expected to be true for certifiers when they want to judge the investment and value generation skills of a PEF. The combination of up-to-date sector knowledge, experience and intuition at a high competence level should facilitate superior fund selection skills. Because such talent is hard to find, it seams realistic that sector specific monopoly certifiers with superior certification skills should evolve; specialization seems the only way for certifiers to skim the surplus generated from certification.

2.4.2 Comparison of Different Types of Certifiers in a Market with Capital Supply Limitations

2.4.2.1 Adjustments to Market Assumptions

In the previous sections I assumed that all PEFs recommended by well respected certifiers get capital from investors. In reality, this is not the case. In addition to the hurdle of finding a qualified certifier, first-time funds face a second limitation that restricts their market entry: At a certain moment in time, there is only a limited amount of capital that investors are willing

31 According to an empirical study of Lossen (2006), specialization of PE funds on a certain financing stage increases investment returns while industry and country specialization do not significantly impact PE fund performance.

to spend on (first-time) private equity funds.[32] In consequence, certifiers, PEFs and investors consider this in their decisions.

General results from the previous sections can be translated to the situation with capital shortage, but several further aspects become relevant. Unless stated differently I consider a polypolistic market environment for the following analysis.

Capital shortage and the capital supply and demand curves I assume that an exogenously driven share s of all funds in the market can be supplied with capital from investors. Thus, every PEF has a probability s to get capital from investors. I restrict the analysis to the interesting case where $s < 0.5$.[33] The share of funds recommended by certifiers is

$$d = p\gamma + (1-p)(1-\gamma),$$

so d is a single fund's probability of getting a certificate. For $s = d$ any certified PEF gets the desired amount of capital from investors. I use the notation $p^* = p^*(\gamma)$ for any realization of p that leads to market clearing at a given γ.

PEFs' and investors' market power is driven by the relationship between capital supply and demand. For illustration purposes, I display the market equilibria graphically in a p-γ-chart. All p-γ combinations which equal the same capital demand make up a curve with the equation

$$\gamma = \frac{d+p-1}{2p-1} \quad \text{for} \quad p > 0.5 \tag{2.13}$$

which has the slope

$$\frac{\partial\gamma}{\partial p} = \frac{1-2d}{(2p-1)^2} \quad \text{for} \quad p > 0.5. \tag{2.14}$$

32 I assume that all PEFs require the same amount of capital to start their fund like in the first part of the analysis. They are not willing to take less and cannot accept more.

33 For $s \geq 0.5$ the capital supply could not be matched completely in the assumed model because certification of more than 50% of funds is not possible (see Appendix 6 on page 240).

For $d < 0.5$ the slope of the hyperbolic curve is always positive. If a certifier is more restrictive in selecting PEFs, i.e., increases p, a higher share of good PEFs γ is required; otherwise the share of certified funds would decrease. Different levels of capital demand give a non-crossing family of curves which is symmetric for p and γ.

The curve of a given capital supply has the same structure as the capital demand curve, i.e.,

$$\gamma = \frac{s+p-1}{2p-1} \quad \text{for} \quad p > 0.5 \tag{2.15}$$

with the slope

$$\frac{\partial \gamma}{\partial p} = \frac{1-2s}{(2p-1)^2} \quad \text{for} \quad p > 0.5. \tag{2.16}$$

with a fixed value of s.

Figure 2.3 shows the capital demand curve family and an example capital supply curve (green color) in a p-γ-chart. The higher s or d are, the further up and left are the respective curves. A rating agency's option space is highlighted in blue and red. The blue line indicates an area where the certified share of funds is greater than the share investors are willing to invest in. The share of certified funds is not sufficient to match capital supply for points on the red line.

Investors' incentive situation The function of the investor's expected return E_I standardized by the fund return A drives his incentive situation:

$$e_I = \frac{E_I}{A} = p\gamma - (1-p)(1-\gamma) = p + \gamma - 1$$

So the indifference curves for a given e_I have the function

$$\gamma = -p + 1 + e_I \tag{2.17}$$

Figure 2.3: Illustration of p-γ combinations that lead to different relationships between capital supply and demand. The symbols used are as explained in the text and in the nomenclature

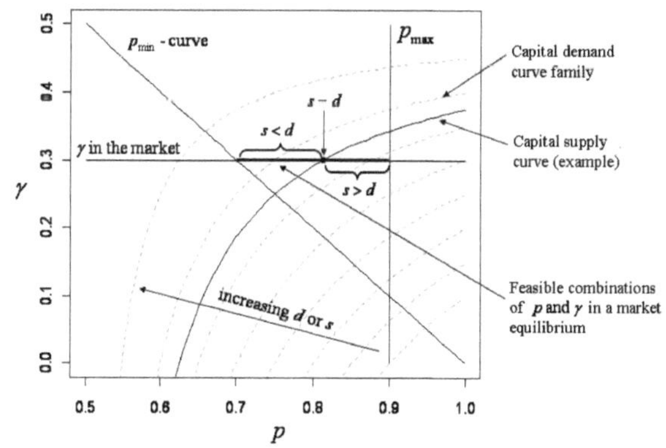

with a slope of -1. The return function e_I is both linear increasing in p and γ as $\frac{\partial e_I}{\partial p} = 1 > 0$ and $\frac{\partial e_I}{\partial \gamma} = 1 > 0$ so investors are always seeking an increase in p and γ.

PEFs' incentive situation In the current model PEFs receive no return because competition drives them down to their reservation utility of zero. Nevertheless they are interested in getting a certificate to start their business. This is why their indifference curves have the same form as the capital demand curves. A PEF's return E_{PEF} standardized by A gives its utility function

$$e_{PEF} = \frac{E_{PEF}}{A} = p\gamma + (1-p)(1-\gamma) = 1 - p - \gamma + 2p\gamma \tag{2.18}$$

which equals

$$\gamma = \frac{e_{PEF} + p - 1}{2p - 1} \quad \text{for} \quad p > 0.5. \tag{2.19}$$

PEFs are better off the more funds are certified, i.e., the higher the level of d. Because $\frac{\partial e_{PEF}}{\partial p} = 2\gamma - 1 < 0$ and $\frac{\partial e_{PEF}}{\partial \gamma} = 2p - 1 > 0$, PEFs want to decrease p and increase γ.

Figure 2.4 illustrates the utility functions of investors and private equity funds.

Figure 2.4: Utility functions of investors and PEFs. The symbols used are as explained in the text and in the nomenclature

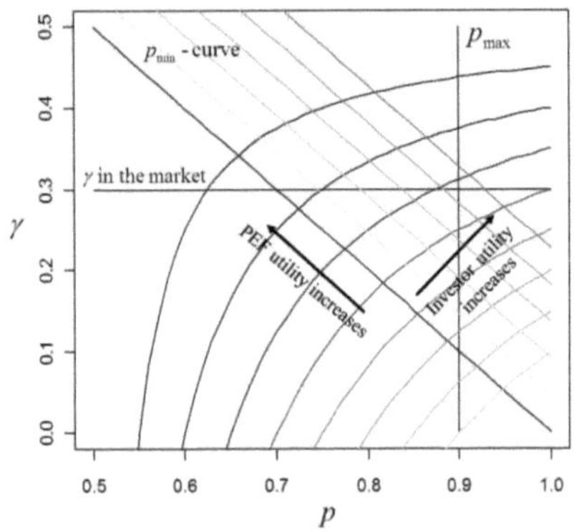

2.4.2.2 Market Equilibrium with a Rating Agency

I assume a market with a given capital supply where rating agencies certify PEFs which are in the process of fundraising. All agencies have the same maximum skill level $p_{max} > 1 - \gamma$.[34] If certifiers only maximize their (fixed) compensation nothing can be said about the level of p they choose.[35] Because RAs rely on reputation, investors and PEFs can steer their behavior (see Sections 2.4.1.2, 2.4.1.2 and 2.4.1.2), rewarding and punishing RAs' reputation to induce the maximization of their respective utility functions:

$$R_{RA} = \begin{cases} R_I^s \cdot (p\gamma - (1-p)(1-\gamma)) - R_{PEF}^s \cdot (p \cdot (1-\gamma) + (1-p) \cdot \gamma) \\ \qquad \text{for} \quad s > p\gamma + (1-p)(1-\gamma) \\ R_I^d \cdot s \cdot \frac{p\gamma - (1-p)(1-\gamma)}{p\gamma + (1-p)(1-\gamma)} - R_{PEF}^d \cdot (1-s) \\ \qquad \text{for} \quad s \leq p\gamma + (1-p)(1-\gamma) \end{cases} \tag{2.20}$$

34 If just one certifier has the skills to realize a higher p_{max} than his competitors, a monopoly would develop as illustrated in Section 2.4.1.3.

35 In reality a certifier would probably select $p = p_{min}$ because this costs him least effort.

In situations of capital over-supply ($s > p\gamma + (1 - p)(1 - \gamma)$) the function is the same as Equation (2.6). When more funds are certified than investors are willing to invest in ($s \leq p\gamma + (1 - p)(1 - \gamma)$), capital supply restrictions limit both reputation rewards and threats from investors and PEFs:

1. Increasing the share of certified funds beyond market clearing does not help PEFs as investors do not spend capital on excess funds.
2. Investors reward an RA for increasing the quality of recommended PEFs; however, the reward is deminished because certifiers blindly restrict the share of the recommendable funds to the level s requested by investors.

Investors and PEFs know when the market is cleared; they reward and punish the certifier differently for fund certification below and above market clearing. I use the nomenclature R_I^d and R_{PEF}^d for reputation reward and threat in capital over-demand and R_I^s and R_{PEF}^s for capital over-supply situations.

The ratio between capital supply and demand in the market determines investors' and PEFs' bargaining power. Rating agencies can face three different possibilities concerning the market clearing skill level $p^*(\gamma)$:

Case 1: $p_{min} \leq p^* \leq p_{max}$ The market clearing level of p is both higher than the minimum level required for investors to be interested in funding PEFs and lower than the certifiers' skill level. To analyze this situation, I consider two parts: First I consider levels of p with $p^* \leq p \leq p_{max}$ and second I analyze levels with $p_{min} \leq p < p^*$. In the first part, all relevant p-γ combinations fulfill the condition $s \leq p\gamma + (1 - p)(1 - \gamma)$. The following Lemma states the rating agency's behavior:

Lemma 10. *In a situation of capital under-supply ($s \leq p\gamma + (1 - p)(1 - \gamma)$) PEFs stop threatening rating agencies' reputation, i.e., $R_{PEF}^d = 0$. When investors select any reputation reward $R_I^d > 0$ rating agencies act in their best interest and maximise p to the market clearing level p^*.*

Proof. In the capital under-supply situation the rating agencies' reputation function is

$$R_{RA} = R_I^d \cdot s \cdot \frac{p\gamma - (1-p)(1-\gamma)}{p\gamma + (1-p)(1-\gamma)} - R_{PEF}^d \cdot (1-s).$$

PEFs cannot induce an RA to minimize p because the reputation function is constant in the PEF's threat

$$\frac{\partial R_{RA}}{\partial R_{PEF}^d} = s - 1.$$

Certifying more funds than necessary for market clearing does not help PEFs as excess certified funds do not get capital from investors. Thus, PEFs can set $R_{PEF}^d = 0$ without affecting rating agencies' behavior; they have no incentive to set a higher R_{PET}^d as this would reduce certifiers' interest to operate in the market.
In consequence, a rating agencies' reputation change just depends on investors reputation reward R_I^d:

$$\frac{\partial R_{RA}}{\partial p} = R_I^d \cdot s \cdot \frac{2 \cdot (\gamma - \gamma^2)}{(p\gamma + (1-p)(1-\gamma))^2} > 0.$$

For any positive reputation reward R_I^d the overall reputation reaches its maximum by maximizing p for a constant γ. The maximum p in the domain of defintion is p^*. □

In the second part of the analysis I consider levels of p with $p^* \leq p \leq p_{max}$, i.e., all relevant p-γ combinations fulfill the condition $s > p\gamma + (1-p)(1-\gamma)$.

Lemma 11. *If capital over-supply induces market power with PEFs ($s > p\gamma + (1-p)(1-\gamma)$), they choose their reputation threat as*

$$R_{PEF}^s > \frac{p + \gamma - 1}{p + \gamma - 2p\gamma} \cdot R_I^s \tag{2.21}$$

for any p-γ combination that restricts the probability to be certified more than necessary for investors to participate. An RA's reputation change R would be negative if it chooses $p > p^*$.

Proof. I start with the assumption that investors select any reputation reward R_I^s in a market with a fixed γ. PEFs observe R_I^s. As stated in Theorem 6, PEFs cannot demand that RAs decrease p because the reputation threat required to make an RA decrease p would cause a negative overall reputation change R_{RA}. A negative reputation change is not a problem in the investigated situation because a rating agency still has the option to certify the market clearing share of funds $d(p^*, \gamma) = s$ or more. Thus, PEFs can completely deter rating agencies from participating in the market in a capital over-supply situation. They simply induce a reputation loss, i.e.,

$$\begin{aligned} R_{RA} &< 0 \\ R_I^s \cdot (p\gamma - (1-p)(1-\gamma)) - R_{PEF}^s \cdot (p \cdot (1-\gamma) + (1-p) \cdot \gamma) &< 0 \\ R_{PEF}^s > \frac{p+\gamma-1}{p+\gamma-2p\gamma} \cdot R_I^s. \end{aligned}$$

□

Lemmas 10 and 11 can be combined into the following Theorem:

Theorem 12. *If a rating agency can select the market clearing level p^*, it will select it. If $p^* < p_{max}$ the overall surplus in the market is below the reachable maximum in the equilibrium.*

Proof. A rating agency balances investors' and PEFs' reputation rewards and punishments in its certification decision. In line with Lemmas 10 and 11 an RA selects p^*.

The function of the average surplus per fund is

$$V = p\gamma \cdot A + (1-p)(1-\gamma) \cdot (-A)$$

which has its maximum in p_{max} because $\frac{\partial V}{\partial p} = A > 0$, i.e., the surplus rises with increasing p. A rating agency that chooses $p^* < p_{max}$ reduces the surplus V generated in the market by $A \cdot (p_{max} - p^*)$. □

Case 2: $p^* < p_{min} < p_{max}$ The situation is characterized by a capital over-supply which is so strong that even the minimum skill level $p_{min} \geq 1 - \gamma$ is not sufficient to meet investors' demand for funds. In this case PEFs have a strong market position. For better illustration consider the following example:

Example 3. *I assume a market with* 100 *PEFs looking for capital; each wants to get €1. The share of good funds is* $\gamma = 0.3$ *and the minimum skill level is* $p_{min} = 0.7$*. Certifiers are limited in their skills at* $p_{max} = 0.9$*. A good fund returns* $A = 1$*, a bad one generates a loss of* -1*.*

First consider a case with capital supply of €46, so 46 funds can get funding and the optimum skill level is $p^* = 0.6$*. At* $p = 0.6$ *investors will lose capital of* $100 \cdot (0.6 \cdot 0.3 - 0.4 \cdot 0.7) = -10$*. Thus, they cannot be interested in* $p^* < p_{min}$*.*

At $p = p_{min}$ *the number of certified PEFs is* $100 \cdot (0.7 \cdot 0.3 + 0.3 \cdot 0.7) = 42$ *each of them creating value of* $1 \cdot \frac{0.7 \cdot 0.3 - 0.3 \cdot 0.7}{0.7 \cdot 0.3 + 0.3 \cdot 0.7} = 0$ *for investors. It follows that investors will earn* 0 *in total. If the certifier increases* $p > 0.7$*, investors earn* $100 \cdot 1 \cdot (p \cdot 0.3 - (1 - p) \cdot 0.7) = 100p - 70$ *which displays a linear increase in p. Consequently, rational investors give certifiers a reputation* R_I *that is linear in p. A PEF has the chance* $p\gamma + (1 - p)(1 - \gamma)$ *to be certified and get capital from investors. The formula is linearly decreasing in p (partial derivation by p equals* $2\gamma - 1 < 0$*). Rationally acting PEFs choose the reputation threat* R_{PEF} *in such a way that the reputation gain from investors for increasing p is totally consumed by the threat except for* p_{min} *because this is the only point that allows PEFs to enter the market.*

Theorem 13. *Rating agencies select the lowest possible* $p = p_{min}$ *in a market with heavy capital over-supply (*$p^* < p_{min} < p_{max}$*). Investors earn nothing.*

Proof. The argumentation is analogous to the proof of Lemma 11: As all market participants know the share of good PEFs γ in the market, they know the minimum level of $p = p_{min} = 1 - \gamma$ for investors' to be interested in certified funds. Investors have a weak bargaining position because their capital supply is too high to be met with a rating agency's skills. PEFs anticipate this and consider the maximum possible supply $s_{max} = d(p_{min}, \gamma)$ as their aspiration level. They threaten RAs' with

$$R^s_{PEF} > \frac{p + \gamma - 1}{p + \gamma - 2p\gamma} \cdot R^s_I$$

as explained in Lemma 11 for any $p > p_{min}$. In consequence, rating agencies certify PEFs benevolently with p_{min}. Investors earn 0 with an average certified fund (see explanations to Theorem 3) which is far below the surplus generated in p_{max} (see Theorem 12). □

Case 3: $p_{min} < p_{max} < p^*$ A high number of PEFs seeking capital are confronted with a low amount of capital supply in the market. This puts investors in a strong position. They can set high requirements for RAs. The contrary is the case for PEFs.

Example 4. *For illustration purposes I reconsider example 3, now assuming that capital of €32 is available in the market and only 32 funds can be funded. This equals a market clearing skill level of $p^* = 0.95$ which is above the maximum skill level $p_{max} = 0.9$. At $p = 0.9$ an RA can recommend 34 funds of which two funds do not get capital from investors. So funds lose their belief in the RA. To avoid this, the RA will blindly select and certify 32 of the 34 identified funds at p_{max} clearing the market. Decreasing p below p_{max} does not make sense for the RA because it couldn't certify more funds. PEFs cannot reward the RA for decreasing p but investors would threaten its reputation for the worse mix of certified PEFs at a lower p. So, investors receive a selection of certified funds which is as good as possible.*

Generally speaking one can state:

Theorem 14. *In a situation of capital shortage where a rating agency's skills are not sufficiently high to reach market clearing, they will choose the maximum skill level possible (p_{max}). Investors' average return per recommended fund is expressed by*

$$\frac{p_{max}\gamma - (1-p_{max})(1-\gamma)}{p_{max}\gamma + (1-p_{max})(1-\gamma)} \cdot A. \tag{2.22}$$

A fund only gets capital with the probability s.

Proof. The main part of the proof was already explained in Example 4: In the situation of extreme capital shortage the whole market power is with investors; PEFs cannot threaten the RA's reputation because certifying more funds than the market clearing level p^* does not change PEFs' probability of entering the market. A rating agency's reputation solely depends on the investors reputation reward; analogous to Lemma 10 PEFs do not threaten an RA ($R^d_{PEF} = 0$) and investors' reputation reward $R^d_I > 0$ makes an RA maximize p (see Lemma 10).

As the maximum share of funds that can get funding is s for $p < p^*$, RAs would certify too many funds even with p_{max} (see Appendix 7 on page 240). A more precise selection of funds is not in the RA's option space, so it blindly issues certificates to the funds it identified as recommendable until the share s of all funds have received a certificate. □

The analysis of cases 1-3 shows that a rating agency can be externally steered by investors and PEFs. The essential assumption in this game is that the certifier exclusively relies on reputation. In this case, the certifier's ability to hide the level of p that he chooses cannot be exploited. So investors and PEFs need not observe p. As long as PEFs and investors have an idea of the share γ in the market and deduct $p_{min} = 1-\gamma$, they can make a rating agency select a p between p_{min} and p_{max} that balances both groups' interests. As long as $p_{min} \leq p^* \leq p_{max}$ capital supply in the market determines the level of p chosen by the RAs.

2.4.2.3 Market Equilibrium with a Placement Agent

The following analysis investigates a placement agent's incentive situation which significantly differs from a rating agency's situation. Discrepancies result from the skills PAs bring to the table and the compensation structure.

A PA does not simply investigate a PEF to identify its quality. He often gets deeply involved in the pre-fundraising phase supporting the fund's concept preparation and finding the right investors that complement the PEF. These activities may partially offset a first-time fund management's weaknesses. The intensive cooperation between PAs and PEFs allows the PA to realize high levels of p, potentially higher than any rating agency is able to reach. In addition, the PA can change the probability of success of PEFs and increase γ in the market, e.g., by supporting the fund management in the fund concept phase. To analyze these effects, I call γ_{min} the (unchanged) market level of γ (because this level can always be reached without supporting PEFs); the maximum level a PA can realize is γ_{max}. A PA may choose any level of γ between γ_{min} and γ_{max}.

PAs receive a fixed share of the funds they raise for a recommended PEF and sometimes get a carried interest in the funds' value generation. If the PA does not recommend and raise a fund he receives nothing for his services.

I apply the same capital restriction rules for the analysis of PAs as in the analysis of rating agencies.

Like rating agencies', placement agents' survival depends on the reputation he has with investors and PEFs. In my model, reputation is the only component that steers a PA's behavior because he is assumed to be committed to long-term survival (see Section 2.4.1.1). Nevertheless it is interesting to analyze under which circumstances short-term gains from a PA's compensation structure would conflict with the goal to maximize reputation. In reality this could cause unintended behavior when PAs consider trading short-term gains against long-term survival.[36] Therefore I analyze

36 I do not explicitly model the trade-off decision but focus my analysis on conflicts of interest between incentives from compensation and reputation.

both determinants - compensation and reputation - separately starting with the effects of reputation.

Incentive effects of reputation Placement agents' reputation functions R_{PA} are the same as the rating agencies' functions:

$$R_{PA} = \begin{cases} R_I^s \cdot (p\gamma - (1-p)(1-\gamma)) - R_{PEF}^s \cdot (p \cdot (1-\gamma) + (1-p) \cdot \gamma) \\ \qquad\qquad\qquad\qquad \text{for} \quad s > p\gamma + (1-p)(1-\gamma) \\ R_I^d \cdot s \cdot \frac{p\gamma-(1-p)(1-\gamma)}{p\gamma+(1-p)(1-\gamma)} - R_{PEF}^d \cdot (1-s) \\ \qquad\qquad\qquad\qquad \text{for} \quad s \leq p\gamma + (1-p)(1-\gamma) \end{cases} \tag{2.23}$$

Theorem 15. *The level of capital supply determines how a placement agent certifies funds when he wants to maximize his reputation:*

1. $s < p_{max}\gamma_{max} + (1-p_{max}) \cdot (1-\gamma_{max})$: *If more funds than necessary for market clearing would be certified at* (p_{max}, γ_{max}), *a PA certifies the market clearing share of funds but thoroughly supports and chooses the funds at* (p_{max}, γ_{max}).
2. $p_{max}\gamma_{max} + (1-p_{max}) \cdot (1-\gamma_{max}) \leq s < p_{min}(\gamma_{max}) \cdot \gamma_{max} + (1-p_{min}(\gamma_{max})) \cdot (1-\gamma_{max})$: *If certifying PEFs at* (p_{max}, γ_{max}) *is not sufficient for market clearing but too many funds would be certified at* $(p_{min}(\gamma_{max}), \gamma_{max})$ *a placement agent recommends the market clearing share of funds, maximizing* γ *up to* γ_{max} *and choosing* $p^*(\gamma_{max})$.
3. $p_{min}(\gamma_{max}) \cdot \gamma_{max} + (1-p_{min}(\gamma_{max})) \cdot (1-\gamma_{max}) \leq s$: *In a market with strong capital over-supply a PA certifies as many funds as possible choosing* $(p_{min}(\gamma_{max}), \gamma_{max})$.

For $\gamma_{max} > \gamma_{min}$ *the surplus generated by a PA is always greater than a rating agency's surplus, but as high as possible just for* $s < p_{max}\gamma_{max} + (1-p_{max})(1-\gamma_{max})$.

Proof. The effect of reputation on a placement agents' behavior is quite similar to rating agencies except for the effect of a change in γ: First assume any level of γ between γ_{min} and γ_{max}. At this (constant) level the certifier realizes

$$p = \begin{cases} p_{min}(\gamma) & \text{for} \quad p^*(\gamma) \leq p_{min}(\gamma) \\ p^*(\gamma) & \text{for} \quad p_{min}(\gamma) < p^*(\gamma) \leq p_{max}(\gamma) \\ p_{max}(\gamma) & \text{for} \quad p_{max}(\gamma) < p^*(\gamma) \end{cases}$$

in line with Theorems 12, 13 and 14. In a situation where $p < p^*$ only investors have market power and induce the PA to maximize p to p^*. The choice of a p-γ combination that restricts the share of certified funds more than necessary for market clearing is not an option for a PA because PEFs threaten the PA with going to a competitor. Only a p-γ combination that allows certification of more than or equal to the market clearing share of funds is feasible, i.e., $p\gamma + (1-p)(1-\gamma) \geq s$. In this case the partial derivations of the reputation Equation (2.23) by p and γ are

$$\frac{\partial R_{PA}}{\partial p} = \frac{2 \cdot (\gamma - \gamma^2)}{(1 - p - \gamma + 2p\gamma)^2} > 0$$

and

$$\frac{\partial R_{PA}}{\partial \gamma} = \frac{2 \cdot (p - p^2)}{(1 - p - \gamma + 2p\gamma)^2} > 0.$$

In consequence, placement agents try to maximize both p and γ in the set of p-γ combinations that allow recommending more funds than the market clearing share. When (p_{max}, γ_{max}) fulfills the condition $p_{max}\gamma_{max} + (1 - p_{max}) \cdot (1 - \gamma_{max}) > s$, a PA selects (p_{max}, γ_{max}); otherwise he chooses γ_{max} and the respective $p^*(\gamma_{max}) < p_{max}$ on the capital supply curve (due to Theorem 12).

The only exception from this behavior occurs in an environment of extreme capital over-supply, where even the maximum share of certified funds at $(p_{min}(\gamma_{max}), \gamma_{max})$ is not sufficient for market clearing. In this case PEFs leave the PAs reputation unpunished for $(p_{min}(\gamma_{max}), \gamma_{max})$ because otherwise PAs don't participate in the market and PEFs don't get capital from investors at all (due to Theorem 13).

The surplus in the market $V = p\gamma \cdot A - (1-p)(1-\gamma) \cdot A$ is positively driven by γ (the partial derivation is $\frac{\partial V}{\partial \gamma} = A > 0$). Because a PA always realizes $\gamma_{max} > \gamma_{min}$, the surplus generated by PAs is always at least $(\gamma_{max} - \gamma_{min}) \cdot A$ higher than the surplus of a rating agency. The maximum possible surplus a PA can realize is the result of the choice (p_{max}, γ_{max}). Because a PA uses this p-γ combination just for $s < p_{max}\gamma_{max} + (1 - p_{max})(1 - \gamma_{max})$ he does not maximize the surplus in all other cases. □

Figure 2.5: Overview of possible equilibrium p-γ combinations for a placement agent and a lead investor. The symbols used are as explained in the text and in the nomenclature

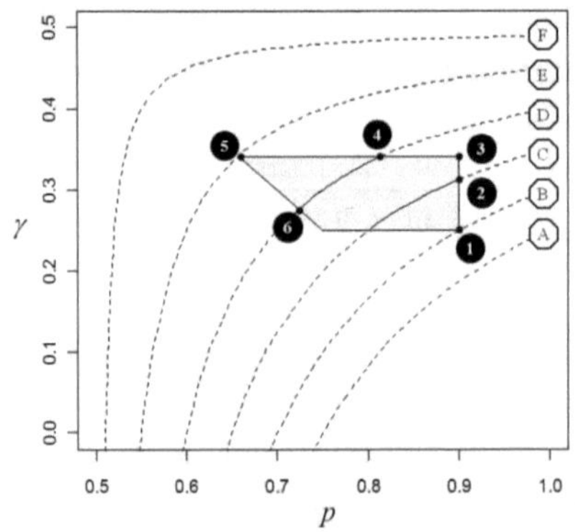

For illustration purposes, I consider figure 2.5 which displays six possible supply curves named A to F. The gray shaded quadrangle represents placement agents' option space for p-γ combinations. Different points are selected in a stable equilibrium depending on the location of the supply curve in the chart: Based on the reputation steering from investors and PEFs (see Theorem 15), placement agents choose point 3 for the capital supply curves A, B and C. For curve D point 4 is optimal and point 5 is the combination of choice for supply levels E and F.

Incentive effects of the compensation structure This paragraph is about the analysis of the contractual terms that potentially cause a conflict of interest between short-term gains from compensation and long-term survival. Therefore I disregard the PAs' dependency on reputation for this analysis and just focus on the incentives their compensation structure induce. I continue with the assumption of a polypolistic market where competition drives PAs' compensation down to their reservation utility. Although the resulting compensation would be 0 in this model, I analyze marginal effects of different compensation components to determine their effects on PAs' behavior.

A PA receives the amount W_{PA} for his services:

$$W_{PA} = \begin{cases} (p\gamma + (1-p)(1-\gamma)) \cdot W_F + p\gamma \cdot W_C \\ \qquad\qquad\qquad \text{for} \quad s > p\gamma + (1-p)(1-\gamma) \\ s \cdot W_F + p\gamma \cdot \frac{s}{p\gamma+(1-p)(1-\gamma)} \cdot W_C \\ \qquad\qquad\qquad \text{for} \quad s \leq p\gamma + (1-p)(1-\gamma) \end{cases} \tag{2.24}$$

$W_F > 0$ equals the fixed amount which a PA gets for his support in the fundraising of the recommended PEFs that get capital from investors.[37]
$W_C \geq 0$ is the carried interest he receives when he detects and recommends a *good* PEF management team.

Theorem 16. *A placement agent who wants to maximize his compensation and does not consider reputation effects decides according to the capital supply in the market and his expected carried interest payments:*

1. *If* $s \leq p_{max}\gamma_{min} + (1-p_{max})(1-\gamma_{min})$:
   Without carried interest payments ($W_C = 0$)*, the PA does not support funds but diligently chooses the funds he certifies, realizing* (p_{max}, γ_{min}).
   With carried interest payments ($W_C > 0$) *the PA supports PEFs as well as possible and diligently selects the funds he certifies, realizing* (p_{max}, γ_{max}).

37 In reality this is by contract a relative share of the volume raised by a PA - usually 2-3%. Because I do not model different fundsizes, W_F can be seen as fixed.

2. *If* $p_{max}\gamma_{min}+(1-p_{max})(1-\gamma_{min}) < s \leq p_{max}\gamma_{max}+(1-p_{max})(1-\gamma_{max})$*:*
Without carried interest payments ($W_C = 0$)*, the PA supports funds up to the market clearing* $\gamma^*(p_{max})$ *and diligently selects the funds he certifies, realizing* $(p_{max}, \gamma^*(p_{max}))$*.*
With a carried interest payment $W_C > 0$ *the PA supports PEFs as well as possible and diligently selects the funds he certifies, realizing* (p_{max}, γ_{max})*.*
3. *If* $p_{max}\gamma_{max} + (1-p_{max})(1-\gamma_{max}) < s$*:*
With low carried interest payments $W_C < W_F \cdot \frac{1-2\gamma}{\gamma}$ *the PA supports PEFs as well as possible (realizing* γ_{max}*) and chooses the market clearing level* $p^*(\gamma_{max})$ *or* $p_{min}(\gamma_{max})$ *if* $p^*(\gamma_{max}) < p_{min}(\gamma_{max})$*.*
With high carried interest payments $W_C \geq W_F \cdot \frac{1-2\gamma}{\gamma}$ *the PA supports PEFs as well as possible and diligently selects the funds he certifies, realizing* (p_{max}, γ_{max})*.*

Proof. To determine the influence of p and γ on the compensation W_{PA}, I partially derive the compensation function by the two variables:

$$\frac{\partial W_{PA}}{\partial p} = \begin{cases} \overbrace{(2\gamma-1)}^{<0} \cdot W_F + \overbrace{\gamma}^{>0} \cdot W_C & \text{for} \quad s > p\gamma + (1-p)(1-\gamma) \\ \frac{\gamma-\gamma^2}{(p\gamma+(1-p)(1-\gamma))^2} \cdot s \cdot W_C > 0 & \text{for} \quad s \leq p\gamma + (1-p)(1-\gamma) \end{cases} \tag{2.25}$$

$$\frac{\partial W_{PA}}{\partial \gamma} = \begin{cases} (2p-1) \cdot W_F + p \cdot W_C > 0 & \text{for} \quad s > p\gamma + (1-p)(1-\gamma) \\ \frac{p-p^2}{(p\gamma+(1-p)(1-\gamma))^2} \cdot s \cdot W_C > 0 & \text{for} \quad s \leq p\gamma + (1-p)(1-\gamma) \end{cases} \tag{2.26}$$

As the partial derivations show, a PA's certification decision critically depends on the carried interest component W_C in his compensation because

$\frac{\partial W_{PA}}{\partial \gamma} = \frac{\partial W_{PA}}{\partial p} = 0$ if $W_C = 0$ and $s \leq p\gamma + (1-p)(1-\gamma)$. A PA without receiving carried interest just maximizes γ to a market clearing level (if this is in his option space); without carried interest payments he tries to reach the highest possible level of γ.

The choice of the PA's quality detection skill p is not as straightforward. With carried interest payments the PA maximizes p for $s \leq p\gamma+(1-p)(1-\gamma)$. For $s > p\gamma + (1-p)(1-\gamma)$ the PA's intentions depend on the ratio between W_F and W_C. The condition for $\frac{\partial W_{PA}}{\partial p} > 0$, i.e., for maximization of p, is

$$\frac{W_C}{W_F} \geq \frac{1-2\gamma}{\gamma};$$

otherwise he tries to minimize selection efforts p in his option space of p-γ combinations.

In addition to the carried interest component in the compensation, the capital supply level s in the market is a key decision criterion. For the assessment of its' influence I consider three cases. All statements are based on the partial derivations of the compensation function (Equations (2.25) and (2.26)).

1. With very low capital supply $s \leq p_{max} \cdot \gamma_{min} + (1-p_{max}) \cdot (1-\gamma_{min})$ all feasible p-γ combinations fulfill the condition $s \leq p\gamma+(1-p)(1-\gamma)$.[38] Therefore, (p_{max}, γ_{max}) is the best choice for a PA that receives carried interest payments because $\frac{\partial W_{PA}}{\partial p} > 0$ and $\frac{\partial W_{PA}}{\partial \gamma} > 0$. Without these payments, he has no incentive to put efforts into certification ($\frac{\partial W_{PA}}{\partial \gamma} = \frac{\partial W_{PA}}{\partial p} = 0$) and would like to recommend as few funds as possible with (p_{max}, γ_{min}).
2. With medium capital restrictions (i.e., $p_{max}\gamma_{min}+(1-p_{max})(1-\gamma_{min}) < s \leq p_{max}\gamma_{max} + (1-p_{max})(1-\gamma_{max})$) the selection of $(p_{max}, \gamma^*(p_{max}))$ is always advantageous (see Appendix 9 on page 241).[39] A further increase of γ from $\gamma^*(p_{max})$ to γ_{max} only makes sense for a PA who

38 The supply level of curve A in Figure 2.5 illustrates this case.
39 The supply level of curve C in Figure 2.5 illustrates this case.

gets carried interest payments because for $\frac{\partial W_{PA}}{\partial \gamma}$ to be positive for $s \leq p\gamma + (1-p)(1-\gamma)$ requires W_C to be positive.

3. Under conditions of high or very high capital supply in the market[40], i.e., $p_{max}\gamma_{max} + (1-p_{max})(1-\gamma_{max}) < s$, the PA maximizes γ to γ_{max} for $s > p\gamma + (1-p)(1-\gamma)$. For $s \leq p\gamma + (1-p)(1-\gamma)$ the PA is either indifferent between diffent levels of γ (for $W_C = 0$) or he also wants to maximize γ (for $W_C > 0$).
   With carried interest payments the PA can increase his compensation by maximizing p to the market clearing level $p^*(\gamma_{max})$ for $s \leq p\gamma + (1-p)(1-\gamma)$. For the analysis if a further increase up to p_{max} yields a positive return for the PA, the partial derivation by p in the area of $s > p\gamma+(1-p)(1-\gamma)$ is relevant. The derivation is positive, i.e., induces maximizing p to p_{max}, if carried interest payments are sufficiently high ($W_C \geq W_F \cdot \frac{1-2\gamma}{\gamma}$). Otherwise the placement agent chooses $(p^*(\gamma_{max}), \gamma_{max})$. If $p^*(\gamma_{max})$ is not in his option space due to minimum requirements from investors for p he chooses $(p_{min}(\gamma_{max}), \gamma_{max})$.

□

From Theorem 16 one can elucidate the effects of a fixed compensation component and carried interest payments on a placement agent's behavior. The fixed compensation - which a PA only receives when he certifies and raises capital for a PEF - makes the PA act in the PEFs' best interest. The carried interest is like a call option that is close to an investor's position, i.e., it is worth something only if the PEF is good. Figure 2.6 illustrates the respective effects.

The comparison of the results from Theorem 15 and 16 reveals that incentives from compensation contracts can conflict with the equilibria induced by reputation concerns. If a PEF wants to avoid this, he has to align incentives:

1. In situations of capital shortage, e.g., supply curves A, B and C in Figure 2.5, reputation motivates a placement agent to maximize his

40 The supply levels of curves D, E and F in Figure 2.5 illustrate this case.

Figure 2.6: Utility function and relevant set of p-γ combinations for placement agents. The symbols used are as explained in the text and in the nomenclature

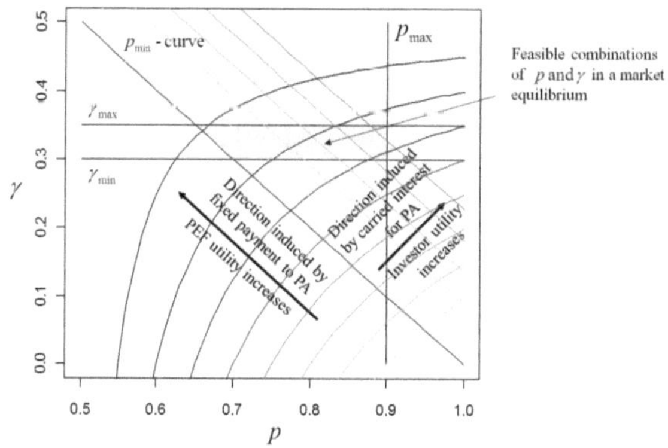

support for the PEFs (γ_{max}) and his selection skills (p_{max}). He will choose point 3 in Figure 2.5. The same is induced with a compensation contract that contains carried interest payments.

2. In a market environment with high capital supply, e.g., supply curves D, E and F, carried interest payments are counterproductive for PEFs. Reputation motivates a placement agent to maximize his support for the PEFs but he will use a lower selection skill than p_{max} to reach market clearing. Sufficiently high carried interest payments to a PA would make the PA act in investors' interest and limit the share of certified funds (realizing point 3 in Figure 2.6). To avoid this, PEFs should not contract carried interest in situations of capital over-supply.

The analysis shows that carried interest payments are critical for the alignment of incentive effects from compensation. The market situation should be considered carefully when deciding upon a compensation contract design for placement agents.

2.4.2.4 Market Equilibrium with a Lead Investor

As explained in Section 2.3.4, lead investors (LI) also play an important role in supporting first-time funds in entering the market. These players have a compensation structure similar to all other investors who spend their capital on a PEF. Just the share of the fund they buy is different - usually they take quite a significant stake in a fund's total committed capital. The high share in the fund gives them a strong incentive to perform detailed checks on the fund management before investing. Moreover, they can get involved in important investment decisions during fundraising, the investment and divestment phases of the fund by participating in the fund investment committee.

I assume that LI involvement in a first-time fund equals an improvement of γ as placement agents can generate. However, an LI's role significantly differs from the one of a PA: An investor usually plays the role of a lead investor only once or a few times in a row, making his dependency on reputation neglectible. Nevertheless, lead investors act like certifiers in the market and take their competitors' behavior into account.

LIs are often involved in important decisions during the fund's lifetime which gives them opportunities to capitalize on the other non-lead investors of the fund. I call the exploiting behavior *collusion* because the fund management team needs to cooperate with the LI for the LI to be able to extract value from other fund investors. Return from collusion may have different sources as explained in paragraph 2.3.4.1.

I investigate the LIs' incentive structure with and without collusion because a PEF management may not want to collude or an LI may have no opportunities to exploit his position.

Lead investors' incentive situation without collusion I consider different types of investors (i.e., lead investors and other non-lead investors) in the analysis. The LI's share $0 < \eta < 1$ in the fund is of special interest. For the first part of the analysis I assume that an LI only gets a return from this share. So his compensation without collusion W_{LI}^{NC} standardized by A is w_{LI}^{NC} with

$$w_{LI}^{NC} = \frac{W_{LI}^{NC}}{A} = \begin{cases} \eta \cdot (p\gamma - (1-p)(1-\gamma)) \\ \qquad\qquad \text{for} \quad s \geq p\gamma + (1-p)(1-\gamma) \\ \eta \cdot (p\gamma - (1-p)(1-\gamma)) \cdot \frac{s}{p\gamma+(1-p)(1-\gamma)} \\ \qquad\qquad \text{for} \quad s < p\gamma + (1-p)(1-\gamma) \end{cases} . \tag{2.27}$$

If there is more than one lead investor in the market, LIs face a free rider problem. This is why I assume that there is a (not clearly specified) marginal incentive for an LI to be the first to certify a fund.[41]

Despite his superior position, an LI is still an investor and thus has the same incentive structure (for graphical illustration see Figure 2.4). The determinants p and γ drive an LI's compensation as:

$$\frac{\partial w_{LI}^{NC}}{\partial p} = \begin{cases} \eta > 0 & \text{for} \quad s \geq p\gamma + (1-p)(1-\gamma) \\ \eta \cdot s \cdot \frac{2\cdot(\gamma-\gamma^2)}{(1-p-\gamma+2p\gamma)^2} > 0 & \text{for} \quad s < p\gamma + (1-p)(1-\gamma) \end{cases} \tag{2.28}$$

$$\frac{\partial w_{LI}^{NC}}{\partial \gamma} = \begin{cases} \eta > 0 & \text{for} \quad s \geq p\gamma + (1-p)(1-\gamma) \\ \eta \cdot s \cdot \frac{2\cdot(p-p^2)}{(1-p-\gamma+2p\gamma)^2} > 0 & \text{for} \quad s < p\gamma + (1-p)(1-\gamma) \end{cases} \tag{2.29}$$

The compensation structure makes the lead investor maximize p and γ, so he fully acts in the best interest of his investor colleagues and neglects negative effects of his decisions on PEFs.[42]

Theorem 17. *Lead investors always support PEFs as reliable as possible and diligently investigate the PEFs' quality choosing* (p_{max}, γ_{max}) *when max-*

41 In reality, preferred investment conditions can avoid the free-rider problem. I do not model preferred conditions, e.g., reduced management fees, because this would just complicate explanations without fundamentally changing conclusions. Preferred investment conditions for an LI raise other investors' participation condition similar to increasing certification costs for a rating agency or a placement agent.

42 This analysis is based on the assumption that a lead investor only fills the certifier role if other investors also invest in the funds he recommends.

imizing their compensation without collusion. PEFs stop mandating lead investors in a market environment with high capital supply ($s > p_{max}\gamma_{max} + (1 - p_{max})(1 - \gamma_{max})$).

Proof. In situations of low capital supply where $s \leq p_{max} \cdot \gamma_{max} + (1 - p_{max}) \cdot (1 - \gamma_{max})$, a lead investor thoroughly supports PEFs and selects the best funds as diligently as possible (see partial derivations in Equations (2.28) and (2.29) for rational). From all funds which he identifies as recommendable, he blindly selects the share s requested by investors. This maximizes his personal compensation and is also in the PEFs best interest, because their chance to get a certificate is independent from the level of γ in this case.

The contrary is true for situations of high capital supply, i.e., $s > p_{max}\gamma_{max} + (1 - p_{max})(1 - \gamma_{max})$: LIs' incentive to support, certify and invest in only the best PEFs (i.e., choose (p_{max}, γ_{max})) restricts the share of certified funds more than necessary for market clearing. Because PEFs cannot obviate this behavior by threatening the LIs' reputation (the LIs do not rely on reputation), they stop mandating LIs for certification purposes. Other certifiers like placement agents offer better chances to get a certificate because they choose $(p^*(\gamma_{max}), \gamma_{max})$ in this case.[43] □

Remark 1. *Lead investors do not regularly act as certifiers. Thus, they have difficulties in communicating their skills p_{max} and γ_{max} to investors and PEFs. This is a problem both for investors and PEFs:*

1. *Investors know that an LI will do his utmost to realize a high return because his own compensation depends proportionally on the fund's success. However, they cannot directly compare the expected return between different LIs' advice and cannot judge the quality of LIs' recommendations.*
2. *PEFs also need to assess p_{max} and γ_{max} to predict if an LI certifies fewer funds than required for market clearing. If they believe that the LI certifies too few funds they prefer another type of certifier.*

43 This statement implicitly assumes that p_{max} and γ_{max} are the same for placement agents and lead investors.

To mitigate their problems, investors and PEFs can inquire into LIs' own investment success in the past. If an LI has been a successful investor this gives a hint of his abilities, though studying the LIs' investment history does not eliminate the information asymmetry problem, but just defers it. One can expect that it is easier to measure an investor's investment success than a new PEF management's investment skills.[44]

Remark 2. *In reality, approaching a (big) investor who may potentially act as a lead investor is always a feasible option for first-time funds. The problem is that any rational investor asks the PEF for a compensation of his investigation and certification efforts. Based on Theorem 17 one can state that - in a market with high capital supply - a PEF should better spend the requested compensation on a different type of certifier, e.g., a placement agent. The PEF's chance to get a certificate is higher with PAs. Moreover, a (big) investor can have non-monetary motives to support a first-time fund, e.g., due to personal relationships between the investor and the fund management. This case is not considered in my model.*

Lead investors' incentive situation with collusion Lead investors may have interests in a PEF other than taking a stake in portfolio companies. By colluding with the fund management they may extract disproportionate returns from the overall value that is generated in the fund at the expense of other (non-lead) investors. I assume that the collusion return is possible

44 If an LI's certificate is a valid signal to investors, the LI does not need a reputation for successful certification but for successful investing.

both from good and bad PEFs alike, so LIs' return with collusion W_{LI}^C standardized by A is:

$$w_{LI}^C = \frac{W_{LI}^C}{A} = \begin{cases} \eta \cdot (p\gamma - (1-p)(1-\gamma)) + \\ k \cdot (1-\eta) \cdot (p\gamma + (1-p)(1-\gamma)) \\ \qquad\qquad\qquad \text{for} \quad s \geq p\gamma + (1-p)(1-\gamma) \\ \eta \cdot s \cdot \frac{p\gamma - (1-p)(1-\gamma)}{p\gamma + (1-p)(1-\gamma)} + k \cdot (1-\eta) \cdot s \\ \qquad\qquad\qquad \text{for} \quad s < p\gamma + (1-p)(1-\gamma) \end{cases} \quad (2.30)$$

where k equals the return from collusion standardized with A.[45]

If, when and how collusion can change the market equilibrium depends on two conditions:

1. Collusion has to have the potential to change a lead investor's certification decision.
2. Investors must be willing to participate in a lead investor's recommended funds while considering offers from competing certifiers in the market.

I assess the fulfillment of the two conditions sequentially:

Theorem 18. *A lead investor who receives a return from collusion, thoroughly supports PEFs, increasing γ as much as possible. In a market environment with low capital supply ($s \leq p_{max}\gamma_{max} + (1 - p_{max})(1 - \gamma_{max})$) the LI maximizes p despite collusion; with high capital supply in the market ($s > p_{max}\gamma_{max} + (1 - p_{max})(1 - \gamma_{max})$) PEFs may motivate the LI to decrease his selection efforts p by offering significant collusion opportunities*

$$k > \frac{\eta}{(1-\eta)(1-2\gamma)}. \quad (2.31)$$

45 This assumption means that a fund's success and an LI's collusion return are uncorrelated. Even if a positive correlation exists between collusion return and fund return, an LI's certification decision will still be equal to the case without collusion. If a negative correlation exists, e.g., just bad PEFs pay a collusion return, an LI will be motivated to certify as many bad funds as possible in any market situation if the expected collusion return passes a certain threshold.

If collusion opportunities are too small, the LI will increase p in the best interest of his investor colleagues. Thus, PEFs do not mandate him.

Proof. I derive the LI's return function w_{LI}^C to determine the effects of changes in p and γ. The derivation by γ is always positive:

$$\frac{\partial w_{LI}^C}{\partial \gamma} = \begin{cases} \eta + (1-\eta) \cdot k \cdot (2p-1) > 0 & \text{for} \quad s \geq p\gamma + (1-p)(1-\gamma) \\ \eta \cdot s \cdot \frac{2 \cdot (p-p^2)}{(1-p-\gamma+2p\gamma)^2} > 0 & \text{for} \quad s < p\gamma + (1-p)(1-\gamma) \end{cases}$$

For $s < p\gamma + (1-p)(1-\gamma)$ the partial derivation for p is

$$\frac{\partial w_{LI}^C}{\partial p} = \eta \cdot s \cdot \frac{2 \cdot (\gamma - \gamma^2)}{(1-p-\gamma+2p\gamma)^2} > 0,$$

which induces a lead investor to increase p up to the market clearing level p^* if this is in his option space. For $s \geq p\gamma + (1-p)(1-\gamma)$ the derivation by p returns

$$\begin{aligned} \frac{\partial w_{LI}^C}{\partial p} &= \eta + (1-\eta) \cdot k \cdot (2\gamma - 1) \\ &\Rightarrow \text{For } k \leq \frac{\eta}{(1-\eta)(1-2\gamma)} \text{ the derivation is } \geq 0. \\ &\Rightarrow \text{For } k > \frac{\eta}{(1-\eta)(1-2\gamma)} \text{ the derivation is } < 0. \end{aligned}$$

Summing up, the partial derivations reveal: If capital supply in the market is low, i.e., $s \leq p_{max}\gamma_{max} + (1-p_{max})(1-\gamma_{max})$, a lead investor maximizes γ beyond the market clearing level $\gamma^*(p_{max})$ to γ_{max} and maximizes p to p_{max}.

High capital supply in the market (i.e., $s > p_{max}\gamma_{max}+(1-p_{max})(1-\gamma_{max})$) enables an LI to capitalize on fund investors if PEFs offer collusion. PEFs have an incentive to offer collusion because this does not affect their own compensation ($E_{PEF} = 0$ anyway)[46]; moreover, it can motivate an LI to

46 In reality, a PEF's compensation usually contains a carried interest which is negatively affected by collusion (Schmidt and Wahrenburg, 2004). I do not consider the PEF's compensation structure in the current model because I argue that a first-time fund is more interested in entering the market at all than in maximizing its compensation. So the compensation effect is not a prior issue.

decrease p which improves the PEFs' chances to enter the market. Thus, PEFs try to promise a collusion return $k > \frac{\eta}{(1-\eta)(1-2\gamma)}$ to the lead investor. If this seems feasible, PEFs mandate the LI; otherwise the LI's incentive to maximize p to p_{max} persists beyond the market clearing level $p^*(\gamma_{max})$. PEFs do not depend on the LI's service (as in the case without collusion possibilities) but may prefer another type of certifier (analogue to Theorem 18).

Assuming a lead investor envisages a collusion return greater than the threshold level $\frac{\eta}{(1-\eta)(1-2\gamma)}$, the incentive to decrease p can motivate the LI to even decrease γ if the return from decreasing p is greater than the loss from decreasing γ. To illustrate this problem graphically, I point to Figure 2.5 with supply curve D. Let's say the lead investor hypothetically starts in point 3. His incentive to decrease p causes a shift to point 4 and potentially even to point 6. He has an incentive to choose point 6, if the compensation is negatively dependent on p for any point on the supply curve. The formula

$$\gamma = \frac{w_{LI}^C + (\eta - k \cdot (1-\eta)) \cdot (1-p)}{\eta + k \cdot (1-\eta)(2p-1)} \tag{2.32}$$

describes the compensation curve family. It equals Equation (2.30) solved by γ. Replacing γ in the supply curve function (Equation (2.15)) by (2.32) and solving for w_{LI}^C returns the function

$$w_{LI}^C = \frac{s\eta + s(1-\eta)k(2p-1) - 2p(1-p)\eta}{2p-1}. \tag{2.33}$$

which equals the compensation function for all points on the supply curve. I partially derive the function to determine the dependency from p:

$$\frac{\partial w_{LI}^C}{\partial p} = \frac{2\eta \cdot (\overbrace{(1-s)}^{>0.5} + \overbrace{2 \cdot (p^2 - p))}^{>-0.5}}{(2p-1)^2} > 0 \text{ for } 0.5 < p < 1 \text{ and } s < 0.5 \tag{2.34}$$

Because the dependency of the LI's compensation from p is always positive, an LI will always try to maximize p and γ on the supply curve independent from collusion; he will select $(p^*(\gamma_{max}),\ \gamma_{max})$ if collusion $k > \frac{\eta}{(1-\eta)(1-2\gamma)}$. □

Deriving Equation (2.32) by γ ($\frac{\partial k}{\partial \gamma} = \frac{2\eta}{(1-\eta)(1-2\gamma)^2} > 0$) reveals that in a market environment with a high share of good PEFs, funds will find it more difficult to motivate a lead investor to decrease p than in a market with a low γ. One reason is that the difference between the market clearing p^* and p_{max} is inversely proportional to γ in the market. So the few additional funds that are certified when decreasing p from p_{max} to p^* must yield the lead investor a higher collusion return than the surplus difference that results from better fund selection.

The partial derivation $\frac{\partial k}{\partial \eta} = \frac{1}{(1-\eta)^2(1-2\gamma)} > 0$ shows that the higher the lead investors share η is in the fund, the more difficult it is for PEFs to motivate the LI to decide in the PEFs' best interest and decrease p. Assuming k has a natural maximum k_{max}, it is not possible for a PEF to efficiently collude with an LI if the share η is too high, i.e.,

$$\eta > \frac{k_{max} \cdot (1 - 2\gamma_{max})}{1 + k_{max} \cdot (1 - 2\gamma_{max})}.$$

Theorem 18 shows that PEFs do not have an incentive to offer collusion to lead investors in a market with low capital supply ($s \leq p_{max}\gamma_{max} + (1 - p_{max})(1 - \gamma_{max})$) because LIs have a self-interest in fully supporting and diligently selecting PEFs with (p_{max}, γ_{max}). This is why I just consider a market with high capital supply ($s > p_{max}\gamma_{max} + (1 - p_{max})(1 - \gamma_{max})$) for the rest of the analysis. In this environment, other investors need to be careful when trusting LIs' recommendations because they can be sure that the LIs receive a return from collusion (otherwise PEFs do not mandate the LIs). From the LIs' share of the fund η investors can even assess the extent of collusion because at least $k = \frac{\eta}{(1-\eta)(1-2\gamma)}$ is required to change a lead investors' behavior in the PEF's interest.

Collusion decreases the (non-lead) investors' expected return because it equals a transfer of the fund surplus to the lead investor. In consequence, investors' alternatives determine if they are willing to accept LIs' recommendations:

1. With no placement agents or rating agencies operating in the market, investors' investment alternatives generate a return of zero (see initial assumptions in Section 2.4.1.1). So investors spend their capital on PEFs recommended by LIs if the expected costs from collusion are lower than the surplus they receive from the LIs' recommendations at $(p^*(\gamma_{max}),\ \gamma_{max})$:

$$k \leq \frac{p^*\gamma_{max} - (1-p^*)(1-\gamma_{max})}{p^*\gamma_{max} + (1-p^*)(1-\gamma_{max})}$$

2. With placement agents operating in the market (having the same p_{max} and γ_{max} as the LIs), investors prefer PAs to LIs. Placement agents realize $(p^*(\gamma_{max}),\ \gamma_{max})$ due to reputation concerns (see Theorem 15) which is the same as lead investors that get sufficient collusion. However, the lead investors require a surplus transfer which is not necessary for the placement agents. Thus, collusion cannot induce lead investors to act in PEFs' interest in a stable equilibrium as investors do not invest in the recommended funds. Investors just consider placement agents' recommendations.

2.5 Conclusion

This study investigates possible ways to overcome barriers for first-time private equity funds to enter the market. To start their first fund, private equity fund managers need to find investors willing to trust them and spend capital on them. The main problem for these managers is information asymmetry with investors due to the fact that they do not have a track record, i.e., a solid basis for investors to judge the fund managers' skills. Studies reveal that the fund managers' skills are a key driver for investment success. Investors' risk in spending capital on a first-time fund is high but necessary especially in growing markets. Intermediaries (in my study called *certifiers*), like placement agents and lead investors, support first-time funds in their fundraising process and in mitigating information asymmetry by signaling the fund managers' quality to the investor community.

In my theoretical model of Section 2.4, I investigate a market with first-time private equity funds where only the minority (less than 50%) of funds have the necessary knowledge and abilities to generate value for investors. In this environment, I assess the incentive situation of placement agents and lead investors as well as rating agencies, who can be seen as a reference group of certifiers. These intermediaries have imperfect but still pre-eminent skills compared to "standard" investors to detect the quality of first-time funds. I assume that intermediaries detect good PEFs' quality with the same probability as they detect bad PEFs' quality. The certifiers maximize their utility which is driven by reputation concerns and/or by their compensation structure. Investors and PEFs try to steer certifiers; investors want them to maximize the investment return and PEFs want the certifier to recommend as many PEFs as possible to increase their chance of entering the market. Both targets may be conflictive.

The goal of my analysis is to determine how the different types of certifiers act under different market conditions and derive suggestions for which partners PEFs should go when entering the market. Major findings are:

1. **Quality threshold of certified private equity fund has to be reached:** Investors expect intermediaries to certify funds with an average quality above a certain threshold; certifiers have to eliminate at least the selection risk with their quality detection skills, i.e., the probability of investing in a bad fund. Certifiers who have worse skills than the minimum skills required lose their reputation over time. Thus, their certificates are no longer accepted by investors.
2. **Certifiers depending on reputation just support investors:** In a market with capital over-supply, certifiers dependent on reputation always act just in the investors' best interest. They restrict fund certification to the best fund quality mix that their skills enable. PEFs

cannot influence a certifier's decision sufficiently to make him certify more funds with a lower average quality. The reputation threat from PEFs which is necessary to make certifiers decrease the quality of recommended funds would have to be so high that the certifiers lose reputation overall and retreat from the market. This is not in the PEFs' best interest; therefore they may stop threatening certifiers in this case.

3. **Superior skilled certifier extracts surplus over second best certifier:** A certifier with superior skills compared to his competitors becomes a "monopolist". He may not extract the full surplus generated in the market but the excess surplus he generates vs. his most skilled competitor. This result is in contrast to Lizzeri (1999) who claims a monopoly certifier can extract the full market surplus. The different conclusions are due to the fact that Lizzeri (1999) assumes perfect certification skills. I assume imperfect skills closer to reality.
4. **Reputation steering induces market clearing:** In a market environment with a fixed amount of capital available for PE investments, the balance between investors' and PEFs' reputation reward and threat drives certifiers' behavior. If a certifier wants to recommend more funds than necessary for market clearing, this would generally be in the PEFs' best interest. However PEFs couldn't gain an advantage because investors don't fund excess certified funds. A certifier does not get reputation from PEFs by excess certification.
   Investors prefer certifiers to recommend fewer funds but at a higher quality. They demand that certifiers push the number of certified funds down towards market clearing. When less than the market clearing level of funds will be certified, the PEFs' reputation threat kicks in and induces the certifier to recommend more funds. So, a certifier dependent on reputation will always recommend a mix of funds with the best possible quality that still leads to market clearing.
5. **Rating agencies cannot keep up with placement agents or lead investors:** Rating agencies are not engaged in the private eq-

uity industry. This fact can be explained using my suggested model. In almost any situation, placement agents and lead investors are superior certification partners than rating agencies both for investors and fund managers. Their ability and incentive to facilitate a PEF's success allows PAs and LIs to a) come closer to a market clearing number of certified funds than a rating agency and b) certify and create a better quality mix of PEFs. Rating agencies just providing a quality statement do not have a chance in the market for private equity investments. This result can also be transfered to other markets. In any situation where a certifying institution which relies on reputation has the opportunity to improve the quality of the goods and services it certifies, it will do so. Market participants depending on certification will listen to certifiers who do not just recommend goods and services but also improve their quality, which is in the certifiers' self-interest.

6. **Placemement agents balance investors' and PEFs' interests in their certification decisions:** Placement agents support first-time private equity funds in the preparation of their funds for market entry both concerning the definition of the fund concept and in selling the concept to investors. The PA's behavior is driven by his long-term goal to survive, i.e., he depends on his reputation among investors and PEFs. Short-term incentives from his compensation structure may conflict with long-term objectives.

   Intensively supporting the proper performance of PEFs is always advantageous for a PA, and this is both in line with investors' and PEFs' interests. When it comes to recommending selected funds, the PA considers mainly the amount of capital available in the market and reputation aspects:

   - In a tight capital market, a PA's option space is quite limited. He selects PEFs as accurate as possible and - if necessary - blindly reduces the share of recommended funds to the share funded by investors. He will support PEFs reliably.

- The PA fully utilizes his abilities in the best interest of the mandating first-time funds if a surplus of capital is in the market. He supports as many PEFs as reliable as possible to maximize their chance to be successful. The PA selects PEFs generously in order to clear the market. In consequence, less than the possible surplus is generated in the market. Because the PA relies on investors' trust he has to ensure a quality standard of the recommended funds. This may cause the market not to be cleared.

A PA wants to maximize his compensation short-term although this is less important than long-term survival. The compensation structure may set short-term incentives not in line with long-term goals. The offer of carried interest payments is critical in this respect. PEFs should agree upon a carried interest payment with PAs in tight capital markets to motivate PAs to support the fund management as reliable as possible while clearing the market. Carried interest payments are counterproductive for PEFs in a market environment with a lot of capital. The payments induce a placement agent to act more in investors' interests and become more selective than necessary in recommending funds.

7. **Lead investors are mainly investors and thus inferior to placement agents as certification partners for first-time funds:** A lead investor does not rely on reputation but just wants to maximize his compensation. Like placement agents, lead investors support PEFs as reliable as possible to increase their chance to be successful. However, lead investors select and recommend funds rigorously independent of the amout of capital in the market. Thus, lead investors are sheet anchors for first-time funds in capital tight market environments. 56% of first-time private equity funds use lead investors. Some fund managers even start their activities just with capital from a lead investor. In any case, lead investors are not first-choice certifiers for PEFs if a lot of capital is available in the market. The certifiers recommend the lowest number of funds with the highest average quality;

so PEFs are not likely to mandate them. To improve first-time funds' chances to enter the market, they can offer to collude with lead investors at other non-lead investors' disadvantage. Because collusion is necessary even when a lot of capital is available, other investors can conclude that lead investors get some form of collusion payments for their fund support. Depending on investors' investment alternatives, non-lead investors accept the lead investors' recommendations anyway if the expected return is higher than the collusion payments. If placement agents operate in the market in parallel, PEFs should stop searching for lead investors at all. Placement agents offer to certify a market-clearing number of funds without expecting collusion payments, a situation which is always to be preferred by the investor community.

3 Study 2: A New Approach to Predict the Performance of Venture Capital Direct Investments

3.1 Introduction

Since the 1980's the asset class PE gained more and more attention worldwide. PE is generally split into VC that focuses on financing start-ups and young companies and Buyout (BO) capital mainly used in leveraged buyouts (LBO) and management buy-outs (MBO) or buy-ins (MBI). This study focuses on VC deals.

Venture Capital bares great chances but also high risks which became especially evident during the end of the last century when massive returns from Internet investments were followed by dramatic losses after the burst of the new economy bubble. In consequence, one may hardly speak of evenly distributed investment returns like in stock or bond markets. In fact, the distribution pattern of the commonly used return measure IRR of private equity investments is highly skewed: A small share (approx. 10%) of investments performs exceptionally well with returns of more than 100% p.a., about 60% have a performance between −99% and +100% and about 30% of investments fall out completely (−100% IRR). The small share of exceptionally high returns has to compensate the significant amount of write-offs for investors to earn a positive return, on average. Some studies, e.g., Ick

(2005) and Phalippou and Gottschalg (2009), even raise doubts that this was achieved in the past on a risk adjusted basis. Thus, finding the right moment in time to invest is a big challenge for every fund manager. He should be highly interested in approaches to foresee his chances and risks for every investment.

Analyzing parameters of investment returns is quite a challenge with the highly skewed distribution pattern of returns. This might be a reason for the relatively little literature in this subject so far compared to the overall literature on PE. An even greater challenge than analyzing return parameters is to predict returns, i.e., foresee future performance as good as possible. This study faces these two challenges and suggests a new approach to analyze and predict investment performance that overcomes deficits in hitherto methodologies.

I consider both macroeconomic and investment specific factors in the prediction approach. My new methodology is a combination of an approach to predict the probability of default of an investment and a method to model the whole distribution pattern of non-write-off investments.

For modeling the defaults, I suggest two statistical methods derived from the credit default modeling literature. In the first method (derived from the *Credit Portfolio View*) I model each investments default probability individually. In the second method (derived from the *Credit Risk Evaluation Model*) I predict shares of write-offs in groups of investments, i.e., in the group of investments that were started in a quarter. For the investigation of the performance of investments that do not need to be written-off, I use a GAMLSS. A GAMLSS allows to analyze parameters of the dispersion of a distribution pattern in parallel to the effects on the average or median of the dispersion[47]. The aspect of dispersion modeling was ignored in the VC literature so far although it is very important when investigating and predicting highly skewed distribution patterns like VC returns.

47 The inventors of GAMLSS use the nomenclature *centrality parameter* for mean or median and *scale parameter* for a parameter describing the dispersion of a distribution pattern. Moreover, *shape parameters* can be modeled in GAMLSS which allow controlling for skewness and kurtosis of a distribution pattern.

To test the power of my approach I analyze data from the CEPRES. CEPRES provided me with highly precise monthly cashflow data on VC direct investments to calculate the investments' return distribution patterns. I statistically investigate the determinants of these patterns to build the basis for my return prediction model. The dominating determinant is the competitive intensity in the VC industry at the time of investing. Moreover, the economic cycle and the cycle of the exit markets (Mergers and Acquisitions (M&A) and Initial Public Offering (IPO)) negatively relate to the investment performance. Worsening IPO market conditions at the time of the first investment in a portfolio company seem to increase the average performance of investments that are not written-off. Moreover, the probability of a "super" investment return increases. Increasing credit spreads – a proxy for the riskiness in the current market environment – seem to induce fund managers to apply a saver investment strategy. This leads to a higher average and a smaller dispersion of return distribution patterns. Early stage investments have a higher default probability, a lower average performance of non-write-offs but a higher dispersion of the performance distribution than later stage investments. Unsurprisingly, I find that investment chances and risks also depend on the industry an investment belongs to. A convex effect on performance is indicated for the investing fund's size, i.e., economies of scale exist for small VC funds but growing too big seems counterproductive.

The current work extends the existing literature on VC investment performance in several aspects: Within a very precise framework, I highlight parameters of chances and risks of VC direct investments alike. I eliminate distortions in the results from other authors that are the consequence of neglecting investment sizes in the analysis of the return distributions. Beyond that, my approach offers a way to predict the performance of VC direct investments. This may support fund managers' and investors' in determining whether a certain moment in time is good for investing. The parameters of my prediction approach are all available before a fund manager takes the first investment decision in a company. Thus, a more reliable prediction of the return distribution pattern for this company should be possible than it can be achieved with other authors' methodologies which require assumptions about the future development of parameters.

The following text starts with a brief overview of existing literature on PE performance determinants. Second, I discuss my suggested performance analysis and prediction approaches. The subsequent description of my methodology shows how I can overcome deficits of other suggested approaches. Next, I use my new methodology to analyze parameters of returns and risks of US VC investments presenting and commenting on the robustness of my results. After the evaluation of the return parameters I use my methodology and predict the performance of historic investments to test the predictive power of the methodology. I compare my results with the results one obtains with the methodology most related to mine from Buchner et al. (2010) and show that my predictions are superior. Before concluding this study, I suggest a new methodology how one can use my investment prediction approach to estimate the performance of a portfolio of VC direct investments, e.g., of a VC fund or fund-of-fund.

3.2 Related Studies

This study relates to several substreams in the PE/VC literature. Since data sources on VC investments became more and more available during the last two decades, increasing research was conducted on the net performance of VC funds and recently also on the gross performance of direct investments of fund managers into portfolio companies.

Literature on the (net) performance of *private equity funds* splits in two kinds of studies. In the first kind, authors like Chen et al. (2002), Jones and Rhodes-Kropf (2003), Kaserer and Diller (2004), Driessen et al. (2009) and Phalippou and Gottschalg (2009) investigate the performance of VC investments descriptively and in comparison to public markets.

Chen et al. (2002) analyzed 148 VC funds from the Venture Economics database between 1960 and 1999. They find that VC funds have an annual arithmetic average return of 45% with a standard deviation of 116%. The geometric average return is estimated to be about 13% which is statistically significantly positive.

Jones and Rhodes-Kropf (2003) investigated the impact of unavoidable principal-agent problems and corresponding idiosyncratic risks on fund returns. They show that VC investments have positive alphas[48] while investors in VC funds earn zero alphas. Fund returns increase with idiosyncratic risks; therefore, more diversified portfolios have lower returns.

In a recent study, Driessen et al. (2009) use a new statistical approach called *General Method of Moments* which works without distribution pattern assumption to assess alpha and beta[49] for 958 VC and BO funds from the Thompson Venture Economics (TVE) database. The authors find a negative Capital Asset Pricing Model (CAPM) alpha of -15% for VC and a beta of 3.21. They identify a high correlation with stock market returns for VC funds.

Another recent study from Phalippou and Gottschalg (2009) concludes that previous studies overstated fund performance, based on a dataset of 1,328 mature private equity funds. The correction for sample biases changes the slight outperformance of funds to a significant underperformance of -3% p.a. vs. the S&P 500 net of fees.

These very diverse results of the mentioned studies raise the question why investors are willing to invest so heavily in VC as they did historically. One of the probable explanations is the lack of transparency on chances and risks in the market. I try to solve this issue – at least partially.

The parameters of fund performance beyond systematic risks are investigated in the second kind of studies on VC funds. Important papers are from Gompers and Lerner (2000), Ljungqvist and Richardson (2003), Ka-

48 A positive alpha indicates an excess return over the market return on a risk adjusted basis.

49 Beta measures the systematic risk of an investment.

plan and Schoar (2005), Kaserer and Diller (2004) and Gottschalg et al. (2004). I draw from these studies to select the parameters of investment returns for my approach.

In their "milestone" study, Gompers and Lerner (2000) highlight the negative impact of high inflows of capital into VC funds on their performance ("money chasing deals" phenomenon).

Kaplan and Schoar (2005) use TVE data from 746 funds and measure market outperformance with the Public Market Equivalent (PME) approach.[50] They report that more experienced funds have a better performance which is persistent. Better performing funds have a higher chance to raise a follow-on fund and new funds are more negatively affected by the market cycle than established funds. They find a concave relationship between fund size and fund return suggesting decreasing returns to scale.

Ljungqvist and Richardson (2003) noticed that the TVE data used by other authors suffers from biases due to self-reporting and accounting problems. Their cash flow data is from one of the largest US PE investors and seems to be more appropriate to analyze performance, although first-time funds and VC funds are underrepresented. Ljungqvist and Richardson (2003) show that fund managers react on changes in competitive conditions in the PE market – mainly concerning competition for deal flow – when timing investments and exits. Their results are central to understanding parameters of capital draw-downs and capital distributions within funds. Although their analysis on investment success parameters points to some dependencies, their performance measure is too simple (they use time-adjusted multiples) to be sufficiently precise to allow performance predictions.

In a study of 200 mature European funds between 1980 and 2003 from the TVE database Kaserer and Diller (2004) find that apart from risks the PE market capital flows, market sentiment and skills of general partners affect fund performance.

50 The PME approach compares a VC investment with a public reference market. Different definitions for PME exist. Commonly, the PME measures how much one has to invest in a reference market to get the same disbursements as with the VC investment. For more details see Kaserer and Diller (2004).

Gottschalg et al. (2004) show that fund performance positively co-varies with business and stock market cycles; returns highly depend on the valuation levels at the time when investments are exited.

On the *direct investment level*, results in literature differ profoundly about the conclusion whether or not VC outperforms public markets. Studies by Quigley and Woodward (2003), Woodward and Hall (2003) and Peng (2001) build venture capital indexes based on a "method-of-moments repeat sales regression" and analyze the correlation of the calculated index with public markets.

Quigley and Woodward (2003) find gross real returns of VC investments of about 5% which are below S&P and NASDAQ returns in the investigated period. Similarly, Moskowitz and Vissing-Jorgensen (2002) ascertain that "early-stage entrepreneurial investments"[51] perform below the market on a risk-adjusted basis.

These conclusions are incompatible with results by Peng (2001) who finds an average geometric return of 55% and a high beta of 4.66 on the NASDAQ for 1987-1999. Woodward and Hall (2003) also find an average performance of VC investments of 20% p.a. outperforming the NASDAQ with an abnormal performance (alpha) of 8.5% p.a. at a beta of 0.86. Results by Cochrane (2005) are more bullish: He analyzed the performance of VC investments in 7,765 companies from the Venture One database correcting for sample selection and identified an annual gross return of 59% and an alpha of 32%.

Such contradictory results reveal a central problem in VC performance investigations: The analysis is not simple and straight forward as in other markets due to selection biases, problems with the measurement of returns and the high skewness of return distribution patterns.

To better cope with these problems, the current study proposes a new analytical approach based upon the performance prediction literature. The

51 This is a very broad definition of VC.

papers mentioned above usually apply methodologies which do not allow serious performance prediction.

So far, performance prediction has mainly been considered on the fund level. Some authors, like Weidig (2002a), Meyer and Weidig (2003), Takahashi and Alexander (2002), de Malherbe (2005) and Buchner et al. (2008) present models to estimate fund cash flows and performance.

Weidig (2002b) suggests an approach to use interim IRR figures of a fund to project the final fund IRR. Meyer and Weidig (2003) propose to calculate fund cashflows based on comparable funds with a similar "internal age" taking into account differences in contribution and disbursement timing of funds. However, both approaches are poor in considering environmental factors influencing investment/divestment timing and performance.

Takahashi and Alexander (2002), de Malherbe (2005) and Buchner et al. (2008) are suited to estimating average capital draw-downs and capital disbursements of funds. These authors explore the risks involved in investment and divestment cashflows separately and elucidate that systematic risks in the funds are time-dependent.

The mentioned models help fund investors to plan their capital requirements for future draw-downs and their reinvestment activities after receiving future capital disbursements from funds. The shortcomings of these approaches are that they do not allow to calculate a detailed prediction of risks and returns of individual investments depending on environmental conditions. However, such results are very interesting for fund managers as well as fund investors when they have to take their investment decisions. A methodology to overcome the shortcomings of existing approaches is the focus of this study.

The paper by Buchner et al. (2010) is closest to my investigation both concerning data and objective. The authors propose a simple model to estimate risk and return distribution patterns of individual investments and combine them into funds or fund-of-funds with a Monte Carlo simulation. However, their model has two weaknesses: First, they use predictor variables which are not available at the time of investing and need to be estimated, e.g., they use the average variation of the Gross Domestic Product (GDP) growth per annum over the entire investment period as parameter. Second, they just model so-called "normal" performers and do not consider dependencies of write-off and extreme performing investments which heavily affect the overall performance of VC. A possible extension to Buchner et al. (2010) is presented by Krohmer and Man (2008) who propose an approach to model write-offs. Their model is very case-specific and hard to generalize.[52]

More details on the methodology of Buchner et al. (2010) and differences between their and my new approach will be explained in Section 3.3.3.1. The integration of my projection methodology for direct investments into their approach to predict fund returns is detailed in Section 3.5.1.

3.3 Methodological Considerations For a New Direct Investment Return Prediction Approach

Researchers in Private Equity face multiple problems when they want to analyze and predict the performance of funds or direct investments. These problems are due to the illiquidity of the market, limited transparency and irregular observations of returns. Especially the fact that actors in the private equity market hardly publish returns and that the PE industry lacks a well functioning secondary market, leads to infrequent and publicly un-

52 Krohmer and Man (2008) investigate different PE stages and industries individually. They try to predict respective default rates in the stages and industries with different regression models. Generalization seems difficult as the authors rely on few observations per regression model making the regressions less robust – especially over time.

observable investment successes and failures. Thus, the key challenges for researchers are:

1. Getting data on funds and direct investments on a sufficiently detailed level including both successful **and** write-off investments,
2. finding the "right" performance measure,
3. finding a good (statistical) approach
   - to model high skewness and kurtosis in investment returns and
   - to correct for potential sample selection, and
4. finding the right parameters affecting performance.

These problems are severe if researchers want to elucidate general dependencies between the returns in private equity and potential return parameters. They are even more severe if one wants to foresee future returns. The following subsections refer to approaches used in existing literature to compensate the inherent problems and present my approach to deal with these issues.

3.3.1 Data Completeness and Representativeness

Until the mid 1990's getting data on private equity investments from public sources was almost impossible. Since then, multiple companies began collecting data on deals of PE funds which constitute a good source for researchers. The main providers are Thompson (with TVE), VentureOne and CEPRES. Many studies (see Section 3.2) were based on TVE, a very comprehensive data collection of worldwide PE funds and respective net cashflows to investors. Other authors like Gompers and Lerner (1997) used data from big institutional investors, e.g., pension funds, for their studies, although this may cause a sample selection bias. The data providing investors might not be able to access very good funds or the dataset can suffer from a bias to better returns if data providing investors have access to very good funds and do not (have to) invest in historically bad performing fund management teams (the later is the case in Gompers and Lerner, 1997).

A data source comparable to TVE both in size and quality, but with more detailed information on direct investment/deal level is the CEPRES database. I use this data source for the current study The CEPRES database contains venture capital, buyout and mezzanine fund investments with respective deal cashflows gross of fees. Thus, the data is unbiased by management fees and carried interest payments of investors to the fund management team. The dataset seems to be quite representative as an analysis shows in Section 3.4.1.1. Nevertheless, because I lack a dataset to compare with I cannot rule out biases caused by funds not recorded in the dataset. Yet, the main concern that unsuccessful individual investments are not present in the database can be ruled out.[53] CEPRES comprehensively records all cashflows between funds and portfolio companies and captures both successful investments and write-offs.

3.3.2 Performance Measurement

Performance measurement in the VC market is more complicated than in other markets due to the 'private' structure of the industry. As VC funds have no legal obligation to publish information, deciphering the performance of an investment is a challenge for any external. Some authors like Gompers et al. (2005) use indicators like the successful exit via IPO or M&A deals as a proxy for (high) performance. Although they state that there is a high correlation between performance and the exit via IPO or sale, this approach is too vague for my purposes. My exact and comprehensive information on monthly cashflows between VC fund and a portfolio company allows me to measure the return of investments more precisely than other researchers can do.

However, even precise cashflow information does not solve the whole problem: As VC investments lack a liquid (secondary) market, one may just estimate the value of an investment to determine its' performance before it is fully realized. Interim investment valuations (net asset values) by fund managers published in their regular reporting to investors may provide a fair estimate of the respective value of an investment in the last few months, but

53 A main concern in other studies is that a dataset is biased due to window dressing activities by GPs or investors who supply the data for the dataset. These activities would cause an upward bias if write-off investments are not or insufficiently recorded.

they are quite skewed in the beginning (Kaserer and Diller, 2004, p. 10f, Kaplan and Schoar, 2005, p. 1799-1801). In some articles, e.g., Kaserer and Diller (2004, 2007), the authors investigate mature funds with the majority or all of their portfolio companies divested to compensate the bias from subjective valuation of portfolio companies. Because my dataset is sufficiently large I am able to be more strict, ruling out these biases completely by just considering fully-realized investments.

In order to measure returns, the private equity industry and the research community have developed a wide range of measures which all have their advantages and disadvantages (see Askar, 2008). In practice, VC funds usually report the performance of their investments and their whole fund in *return multiples* like Total Value to Paid-in Capital or in *IRR*. Academics added return measures like *MIRR* and the related *Profitability Index*, as well as the *Public Market Equivalent PME* and the related *Static Spread* approach in recent years.

The PME and Static Spread are relative return measures just indicating an out- or underperformance versus a public (stock or bond) index. These approaches are not useful for my study as I want to predict *absolute* returns which are interesting for fund managers and investors. Moreover, return multiples are useless for my purposes as well as they completely neglect the time-value of money and opportunity costs.[54]

The IRR considers the time value of money and is as such a widely used and well respected measure. This is why I focus on the IRR for the current study and analyze IRR figures calculated from cashflows between the first investment decision and the final exit from an investment. The IRR is defined as the average return of the invested capital over the investment

54 For an extensive review of performance measures in the PE industry see Diller (2007) and Askar (2008). Askar (2008) shows that the comparison of VC funds based on different return figures leads to different preferences.

period. For fully realized investments (i.e., investments without a remaining net asset value) it is calculated by solving

$$\sum_{t=0}^{T} \frac{c_t}{(1+IRR)^t} = 0$$

for IRR. Here, c_t are the (positive or negative) cashflows in period t with T standing for the investment maturity.

The IRR is a value weighted performance figure which is affected by the time structure of the cashflows. Because VC fund managers may influence the timing of investments – this should be one of their key skills – the IRR is a suitable measure. Despite its common usage the IRR has several downsides (see Diller, 2007, p. 91-94) which I can partially but not fully mitigate. In consequence, results always have to be treated with some care by the interested party:

One problem of the measure is that the mean IRR of individual investments does not equal the IRR of the pooled cashflows of these investments.[55] The reason is that this mean IRR does not consider investment sizes and durations. Moreover, IRRs take an unrealistic reinvestment assumption, i.e., one implicitly assumes in the calculation that disbursements from the investment are reinvested with a return that equals the IRR itself. I use three approaches to mitigate these problems:

1. Instead of just analyzing the average IRR of all investments, I investigate the whole distribution pattern (including percentiles) of the IRRs of all individual investments in a cross-section analysis.
2. Neglecting the different invested amounts when calculating percentiles of the IRR distributions, implies neglecting the different weights of the observations in the calculation. This problem is often unconsidered in academic literature but may have significant effects on drawn conclusions because investment sizes differ substantially. Phalippou and Zollo (2005) point to the negative effect of the weighting correc-

55 Cashflow pooling means summing the cashflows of all investments up per time unit and calculating the return of the sum of the cashflows.

tion on performance on the fund level and find substantial effects on the average performance. I explicitly consider investment sizes in the weighting of the observations in my regression analysis. Therefore, I normalize investment size on the number of observations, i.e., my weighting just changes the relative importance of an observation, the total weighting equals the sample size.

3. To consider possible biases from the implicit opportunity cost assumption in the IRR calculation, I cross-check the results obtained with the IRR as performance measure with the MIRR which was originally suggested for PE by Dorsey (2000, p. 19f).[56] The MIRR assumes that cashflows not invested in the beginning of an investment and cashflows returned before the end of an investment are discounted/accumulated with a suitable investment rate to the start/end of the holding period.[57] The MIRR is calculated as follows:

$$MIRR = \sqrt[T]{\frac{\sum_{t=0}^{T} CFO_t \cdot (1+r)^{T-t}}{\sum_{t=0}^{T} CFI_t \cdot (1+r)^{-t}}} \tag{3.1}$$

CFI_t represents the cashflows into the portfolio company in time t while CFO_t stands for the cashflows back to the fund. r is the market return resp. investment rate and T the total investment duration. For the market return I take the standard assumption of a constant rate equaling the average annual return of a comparable public stock market index (e.g., Lossen, 2006). I use the annual return of the NASDAQ in the time frame of my observations because the index mainly consists of high-tech companies which are also the focus of VC investing. Thus, a quite similar risk profile is considered.[58]

56 The problems of IRR are known for a long time; Jean (1970) discusses the concept. – Brigham and Gapenski (1997) present a definition and a simple methodology of calculating a unique MIRR for any series of cashflows. For a complete discussion of MIRR methodologies please refer to McDaniel et al. (1988).

57 Dorsey (2000) does not suggest an appropriate rate; he just refers to Brigham and Gapenski (1997) who propose to use the cost of capital as the reinvestment rate.

58 The choice of the reinvestment rate is arbitrary. Every investor may choose his individual rate depending on his investment policy. – Besides from taking the annual return of the NASDAQ, I also tested

3.3.3 Methodology Choice

3.3.3.1 Shortcomings of Existing Methodologies to Analyze Venture Capital Returns

Historically academic researchers in private equity focused on the analysis of parameters of the performance of whole funds due to data limitations. Because of diversification effects the very heterogenous returns[59] of single investments average out in a VC fund, leading to a more normal return distribution pattern than single deals reveal (Weidig and Mathonet, 2004, p. 3); though some heterogeneity remains.

Most older studies use simple *Ordinary Least Square (OLS)* regression to analyze performance. This approach relies on strict assumptions concerning the distribution of regression residuals to allow inference analysis. Central OLS regression assumptions are homogeneity and normal distribution of residuals which are both doubtful due to the high skewness of the direct investment returns constituting the fund returns.[60] To correct for the undesired effects of heteroscedasticity in regression residuals on the conclusions drawn from regressions, most authors use robust estimators for the regressors' standard errors in their specifications.

In a more recent study from Kaserer and Diller (2007) who applied a *Weighted Least Square (WLS)* approach - which is one way to really consider heterogeneity of residuals - the authors obtained an R^2 of 39.1%; this is far more than former studies showed without heterogeneity modeling.

Although robust estimators and weighted regression are both useful approaches in the determination of parameters of investment performance,

other indices like the Dow Jones and the S&P 500 but didn't find major differences in the structure of the results of my investigation.

59 Yung (2008) explains the high level of heterogeneity in a theoretical model. He concludes that the entry of low-quality entrepreneurs and less efficient fund managers cause the "cross-sectional variation in long-run returns [to] be higher within cohorts of firms that were funded during hot markets" than during cold times (Yung, 2008, p. 14).

60 In big funds with many investments the central limit theorem applies, but this is hardly the case in "normal" funds with less than 20 investments. This complication is confirmed by Kaserer and Diller (2007, p. 15f) who show that heteroscedasticity is a problem in the analysis of PE funds.

they are less supportive when one wants to develop a prediction model because the later requires explicitly specifying the heterogeneity.[61]

On the direct investment level - which is the focus of my study - the heterogeneity problem is severe; the high share of investments which are written off completely and the significant amount of investments with exceptionally high investment returns prohibit standard statistical methods like OLS regression. Results are doubtful even with robust standard error estimates. This is why researchers were looking for ways to explicitly consider or avoid the heterogeneity in their analysis. For example, Cumming and Walz (2004) and Lossen (2006) take the natural logarithm of IRRs to limit heterogeneity effects on their OLS regressions.

An approach to get rid of heterogeneity is suggested by Buchner et al. (2010). The authors neglect write-off investments and "exceptionally" high IRRs (>100%) in their regression analysis, and just investigate and predict "normal" performing investments. They are able to explain 21.8% of variation in their data applying an OLS approach. The key problem in their methodology is that the authors are not able to investigate what leads to write-offs and what to exceptionally high performance. This is important, considering the finding of Cochrane (2005, p. 6) that not the "normal" performers drive the average return of Venture Capital, but the high volatility in returns, i.e., write-offs and exceptially high performers. Buchner et al. (2010) argue that non-normal-performing investments are more driven by mere chance than by the economic environment. My analysis indicate the contrary (see Section 3.4.2). Krohmer and Man (2008) confirm the dependency of the default risk of PE investments on macroeconomic parameters.

61 Robust estimators are merely a way to mitigate the undesirable effects of heterogeneity of regression residuals when determining the statistical significance of performance parameters; they do not allow modeling of heterogeneity. - In WLS regressions the heterogeneity is modeled but parameters driving the heterogeneity cannot be specified. However, these parameters are required for a useful prediction of performance distribution patterns.

3.3.3.2 A New Methodology to Analyze Venture Capital Direct Investment Return Distribution Patterns

As mentioned above, current approaches to analyze and predict the performance of VC direct investments are unable to evaluate the whole picture. To meet this challenge, one has to find a suitable methodology to analyze and predict

1. the share of investments which fall out completely and
2. the complete return distribution pattern of investments which are not written off, i.e., to identify what drives the mean/median, the variance and the shape of this distribution.

I present separate approaches for both items which may be used in a combination to predict the whole return distribution pattern of investments at a certain moment in time. Figure 3.1 exemplarily shows the highly skewed and kurtotic distribution of IRRs. Moreover, it illustrates which parts of the distribution pattern are investigated with the different methodologies which are detailed in the following sections.

Analysis of the Share of Write-offs Identifying the probability of a total write-off in a collection of VC investments is analogue to modeling default probabilities in a credit portfolio. One may follow either a very simple approach assuming a constant default rate over time for all investments (this is the approach used by Buchner et al., 2010) or one may try to identify macroeconomic factors leading to changes in the share of defaults. This is relevant because the default rate varies between roughly 10% and 50% over time (see e.g., Figure 3.4). I explicitly investigate the influence of macroeconomic parameters and develop two approaches in analogy to concepts from the credit default literature.

Figure 3.1: Example Distribution Pattern of Investment IRRs and Approaches Applied to Predict Different Parts of the Distribution Pattern in the Current Study

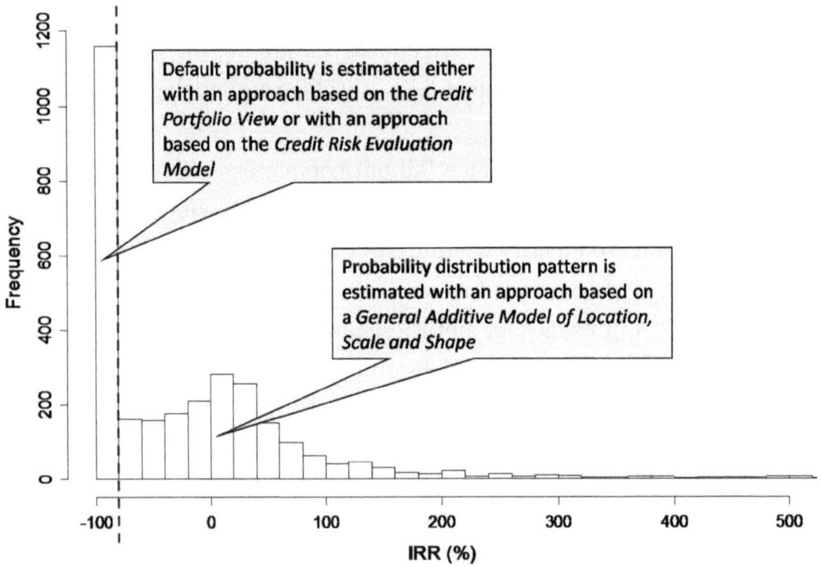

Credit Portfolio View The first approach that relates default rates to macroeconomic conditions is derived from a concept called *Credit Portfolio View (CPV).*[62]

Central to the Credit Portfolio View (CPV) is a logistic regression (Logit model) used to identify parameters of the probability of default π_i of an *individual* investment $i = 1, ..., n$ with n equaling the total number of investments. Therefore, I define the dummy variables w_i for the investments which are 1 if an investment was written off completely and 0 otherwise. In analogy to the standard LOGIT model I assume $\pi_i = \mathrm{P}(w_i = 1|x_i)$ with $x_i = (x_{i,1}, \ldots, x_{i,k})$ equaling a vector of k covariates for each investment i which potentially influence the default probability.

62 Wehrspohn, 2002 gives an overview on the concept for credit default modeling; for an application to VC refer to Krohmer and Man, 2008.

The logarithm of the odds ratio $\frac{\pi_i}{1-\pi_i}$ is regressed on the covariates

$$\ln\left(\frac{\pi_i}{1-\pi_i}\right) = \beta_\pi^{'} x_i = \beta_{\pi,0} + \beta_{\pi,1} x_{i,1} + \ldots + \beta_{\pi,k} x_{i,k}. \tag{3.2}$$

Here, $\beta_\pi^{'} = (\beta_{\pi,0}, \ldots, \beta_{\pi,k})^{'}$ represents the vector of the intensity of the influence of the covariates.

Based on equation (3.2) one can calculate the expected individual default probability $\mathrm{E}(\pi_i)$ of an investment i as

$$\mathrm{E}(\pi_i) = \frac{1}{1 + e^{-(\beta_{\pi,0} + \beta_{\pi,1} x_{i,1} + \ldots + \beta_{\pi,k} x_{i,k})}}. \tag{3.3}$$

Credit Risk Evaluation Model The *Credit Risk Evaluation Model (CREM)* is used to predict the credit default probability in a *group* of credit contracts in dependency from (macroeconomic) factors (Wehrspohn, 2003). The approach contrasts with the CPV approach because default probabilities of credit contracts are not modeled individually. Instead, one assumes that the probabilities are equal across a group of investments but differ between the groups. According to Wehrspohn (2003), the CREM has the advantage that an underestimation bias intrinsic in the CPV method is avoided. Its disadvantage are increased information requirements in contrast to the CPV, as the number of observations per group has to be sufficiently large to get meaningful default rates.

I transfer the concept from credit contracts to VC investments. The default probability δ_j of a group j of investments, i.e., the share of investments that falls out completely, may be modeled with any methodology that fits the data well. A Tobit model (Tobin, 1958) could be applied because the default rate in a group is generally restricted to values between 0 and 1. However, I use a more simple OLS regression approach as I observe that the default rates are hardly impacted by these restrictions.[63] Thus, I esti-

63 I tested the Tobit regression approach but didn't find important differences to the simpler OLS regression methodology.

Figure 3.2: Illustration how CPV Based Approach and CREM Based Approach are Used to Estimate the Default Probability of an Investment

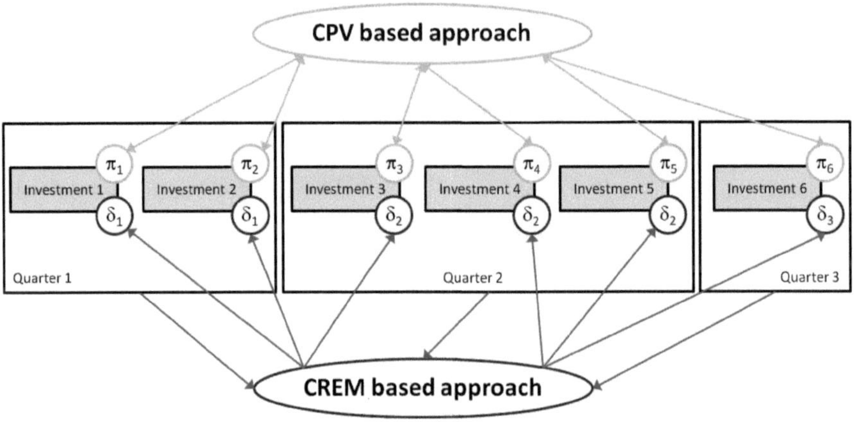

mate the default probability δ_j of all investments in a group j with the OLS regression model

$$\delta_j = \beta_\delta' z_j = \beta_{\delta,0} + \beta_{\delta,1} z_{j,1} + \ldots + \beta_{\delta,m} z_{j,m} + \epsilon_j. \tag{3.4}$$

$z_j = (z_{j,1}, \ldots, z_{j,m})$ equals the vector of the m covariates for the group j and $\beta_\delta' = (\beta_{\delta,0}, \ldots, \beta_{\delta,m})'$ represents the vector of the intensity of the influence of the covariates. ϵ_j is the residual of the OLS regression which is assumed to be normal distributed with an expected value of 0 and a constant variance. From equation (3.4) the expected default probability $\mathrm{E}(\delta_j)$ of all investments in the group j can be derived as

$$\mathrm{E}(\delta_j) = \beta_{\delta,0} + \beta_{\delta,1} x_{1,j} + \ldots + \beta_{\delta,m} x_{m,j}. \tag{3.5}$$

I test the power of both methodologies, the CPV based approach and the CREM, to forecast the default probability of an investment. Figure 3.2 illustrates the differences between the methodologies. I use quarterly data for the CREM based calculations, i.e., every quarter is one observation. A further split of the available observations, e.g., into different investment stages or industries, was envisaged but not applied as this would lead to an insufficient number of investments per observation unit. I use similar explanatory variables (also called covariates) in both calculations. The covariates which I apply in the CPV based approach are aggregated to a quarterly level (if possible) for the CREM based analysis. Some covariates are neglected because an aggregation to quarters is not meaningful.

Analysis of Probability Distribution Pattern of Non-write-off Investments Some authors, e.g., Cochrane (2005) and Buchner et al. (2010), have stated that the return distribution pattern of non-write-off VC investments measured in IRR is very different from a normal distribution. This causes problems in regression analysis.[64] Cochrane (2005) ascertains that a logarithmic data transformation of the return measure IRR can be used to better approximate normality. This approach works well for the identification of return parameters but is insufficient for a good return prediction.

In order to predict the whole return distribution pattern one requires a methodology which allows for identifying parameters that describe all quantiles of the distribution.[65] Generally, two methodologies seem applicable: Quantile regression (Koenker and Hallock, 2001) and General Additive Models of Location Scale and Shape (GAMLSS). I tested both methodologies and chose GAMLSS for this investigation due to better modeling capabilities for my purposes.

GAMLSS were recently introduced by Rigby and Stasinopoulos (2001, 2005) and Akantziliotou et al. (2002).[66] Their methodology can be described as a regression type model to identify parameters of the whole distribution pattern of a dependent variable - including distribution quantiles. Key re-

64 To be precise, a normal distribution is just required for the residuals of the regression to allow inference analysis. My analysis reveals that this has to be doubted in the analysis of IRRs as well.

65 Standard OLS regression just identifies parameters of the distribution mean.

66 The inventors extended regression approaches like generalized linear models (GLM Nelder and Wedderburn, 1972) and generalized additive models (GAM Hastie and Tibshirani, 1993) increasing flexibility.

quirement to a GAMLSS is a parametric distribution assumption for the dependent variable. Based on this assumption, the GAMLSS allows modeling all parameters of the distribution function (in my case especially the centrality and the dispersion parameters) in parallel as linear functions of covariates.[67,68]

Finding the right distribution assumption is key to getting a high quality fit of the GAMLSS to the dependent variable. My analysis indicated that the IRR and the MIRR distribution patterns are both highly skewed and kurtotic in comparison to the normal distribution. A Box-Cox transformation[69] – which is a common way to eliminate skewness in distribution patterns – is not sufficient to end up with the required normal distributed residuals in the regressions. Thus, I tested the fit of a broad set of (skewed and kurtotic) distributions to the IRR and MIRR distribution patterns.[70] The Box-Cox Power Exponential (BCPE) distribution[71] fits excellent to the probability distribution patterns of $\log(100\% + IRR)$ and $\log(100\% + MIRR)$.[72] This is why I illustrate my statistical results assuming a BCPE distribution for the logarithms of IRR and MIRR as dependent variables.

The BCPE distribution was developed by Rigby and Stasinopoulos (2004). It is a combination of the Box-Cox concept for skewness modeling (Box and Cox, 1964) and a power exponential distribution[73] which allows leptokur-

67 I do not explain in detail how the fitting in the GAMLSS works. An interested reader of this article can refer to www.gamlss.com and Rigby and Stasinopoulos (2001, 2005) as well as Akantziliotou et al. (2002) for more details.

68 GAMLSS are rarely used in econometric analysis so far. According to my knowledge, these methods are not yet implemented in commercial software like *STATA* or *SPSS*. I employed the open source software package *R* commonly used by academic statisticians with the library *gamlss*. The general framework in *R* allows fitting almost any distribution pattern of the response variable and model respective parameters describing the distribution pattern.

69 A Box-Cox data transformation methodologically includes a logarithmic data transformation as a special case for the transformation parameter $\lambda = 0$.

70 My analysis included among others the *Inverse Gaussian* family, *Weibull*, *Gamma*, *Reverse Gumbel* and *Johnson's SU* distributions. To choose the best fitting distribution I used the common model selection approach based on the criteria Akaike Information Criterion (AIC) and Schwarz Criterion (SBC).

71 I use a truncated version of the BCPE distribution as values below -100% IRR are not possible.

72 It is necessary to add 100% to the IRR/MIRR values to enable logarithmizing.

73 The power exponential distribution is also called Subbotin or Box-Tiao or generalized error distribution.

tosis and platykurtosis modeling (Rigby and Stasinopoulos, 2004). The distribution function has four parameters that may all be parameterized in a GAMLSS: μ is the median (also called location parameter), σ (also called scale/dispersion parameter) is approximately the coefficient of variation (for small σ and moderate $\nu > 0$), ν controls the skewness and τ the kurtosis of the distribution.

When fitting the BCPE distribution to the return distribution patterns, my GAMLSS model requires o independent observations y_l with $l = 1, \ldots, o$ and assumes

$$y_l|(\mu_l, \sigma_l, \nu_l, \tau_l) \sim \text{BCPE}(\mu_l, \sigma_l, \nu_l, \tau_l), \tag{3.6}$$

i.e., the realizations of investment returns follow a BCPE distribution with the distribution parameters μ_l, σ_l, ν_l and τ_l. I do not estimate ν_l and τ_l in dependency from explanatory variables although this would be theoretically possible in a GAMLSS. Rationale is that a) an overfitting could occur and b) my analysis revealed that these parameters are fairly constant. Thus, I just model the two distribution parameters μ_l and σ_l as functions of covariates, i.e.,

$$\mu_l = \beta_\mu^{'} x_l = \beta_{\mu,0} + \beta_{\mu,1} x_{l,1} + \ldots + \beta_{\mu,k} x_{l,k} \tag{3.7}$$

$$\log(\sigma_l) = \beta_\sigma^{'} x_l = \beta_{\sigma,0} + \beta_{\sigma,1} x_{l,1} + \ldots + \beta_{\sigma,k} x_{l,k}. \tag{3.8}$$

In these equations x_l is a vector of the fixed and known values of the covariates for an investment (basically the same as in the CPV calculations). $\beta_\mu^{'}$ and $\beta_\sigma^{'}$ equal the parameter vectors.
The distribution parameters ν_l and τ_l are modeled as fixed values, i.e.,

$$\nu_l = \nu_{Fix} \tag{3.9}$$

$$\log(\tau_l) = \tau_{Fix}. \tag{3.10}$$

GAMLSS allows to specify different link functions for the distribution parameters. I use an identity link for μ_l and ν_l, i.e., apply no modification to the original terms, and a logarithmic link for σ_l and τ_l as equations 3.8 and 3.10 show.[74]

3.3.3.3 Deriving a Prediction Methodology from the Analysis Approach

The combination of the CPV based model or the CREM based model for default rates with the GAMLSS model for the return distribution pattern of non-write-off investments allows predicting a new investment's whole return distribution pattern.

First, one has to estimate the default rate for a new investment s. In case the CPV based model is employed, the expected default probability of every individual investment may be calculated based on the fixed covariate vector $x_s = (x_{s,1}, ..., x_{s,k})$ for the investment using equation 3.3. To get the default rate of a group of investments one calculates the weighted average (weighting is necessary to control for different investment sizes).

If the CREM method is applied, the default rate of an investment is defined by the (average) default rate of the group it belongs to. To calculate the group specific default rate equation 3.5 may be used. Aggregation to bigger groups (e.g., quarters to years) works with weighted averages in analogy to the CPV based case.

To predict the non-default distribution pattern, the values of the covariate vector x_s has to be multiplied with the GAMLSS parameter vectors β'_μ and β'_σ which define the general relations between the covariates x and location / scale parameters μ and σ. The skewness and kurtosis parameters in the

74 This is the standard approach for fitting a BCPE distribution in the *gamlss* package.

GAMLSS model are fixed. Thus, the respective formula to calculate the expected distribution parameters are:

$$\mu_s = \beta_{\mu,0} + \beta_{\mu,1} x_{s,1} + \ldots + \beta_{\mu,k} x_{s,k} \tag{3.11}$$

$$\sigma_s = e^{\beta_{\sigma,0} + \beta_{\sigma,1} x_{s,1} + \ldots + \beta_{\sigma,k} x_{s,k}} \tag{3.12}$$

$$\nu_s = \nu_{Fix} \tag{3.13}$$

$$\tau_s = e^{\tau_{Fix}} \tag{3.14}$$

With the values for μ_s, σ_s, ν_s and τ_s, the BCPE distribution is fully specified and allows calculation of distribution quantiles.[75] To get the distribution pattern of a group of investments one has to calculate a weighted average distribution.[76]

Finally, the combined performance distribution function $\mathrm{F}_s(p)$ with p equaling the performance measure IRR resp. MIRR needs to join the default probability and the BCPE distribution function:

$$\mathrm{F}_{s,CPV}(p) = \begin{cases} E(\pi_s) & \text{for } p = -100\% \\ E(\pi_s) + (1 - E(\pi_s)) \cdot \mathrm{BCPE}(\log(100\% + p) | \mu_s, \sigma_s, \nu_s, \tau_s) & \\ & \text{for } p > -100\% \end{cases} \tag{3.15}$$

for the CPV based approach and

$$\mathrm{F}_{s,CREM}(p) = \begin{cases} E(\delta_q) & \text{for } p = -100\% \\ E(\delta_q) + (1 - E(\delta_q)) \cdot \mathrm{BCPE}(\log(100\% + p) | \mu_s, \sigma_s, \nu_s, \tau_s) & \\ & \text{for } p > -100\% \end{cases} \tag{3.16}$$

75 A Monte-Carlo simulation extending an OLS-model as suggested by Buchner et al. (2010) to solve the prediction problem is not necessary in this calculation.

76 This approach is applied in the test of my model in Section 3.4.3.

for the CREM based approach. In the CPV case $E(\pi_s)$ is the expected default probability of the investment s. In the CREM case $E(\delta_q)$ is the expected default probability of the group q to which the investment s belongs. $\text{BCPE}(\log(100\% + p)|\mu_s, \sigma_s, \nu_s, \tau_s)$ stands for the value of the BCPE distribution at $\log(100\% + p)$ given the distribution parameters μ_s, σ_s, ν_s and τ_s.

3.3.4 Parameters and Indicators of Investment Performance

The methodology described above requires covariates x to conduct the analysis and prediction of investment performance. Different factors influence the performance of a VC investment from the first investment decision to the final exit and may be used as covariates. For my test of the new methodology, I stick to explanatory variables that are either known at the time just prior to the first investment or easy to estimate/foresee. Although, this makes the analysis quite challenging I don't want to use parameters in my analysis which describe the environment during an investment like Buchner et al. (2010) do. This would require me to guess or estimate the parameters which are used in a second step to predict the performance distribution of an investment. With the parameters being a guess, the prediction becomes a guess as well.

I derive potential performance parameters from the factors considered in VC fund managers' due diligence process when determining a company's value. Depending on the available data, a fund manager or investor, who wants to make predictions, may be able to use further indicators to develop his individual prediction model.

3.3.4.1 Value Creation Potential and Risk Measures

My first set of explanatory variables measures the value creation potential and respective risks involved in a VC investment. An important parameter of the success of a company is the overall economic situation in its' target market. If an investor is able to buy a company or invest in a venture when the market hits its low, he can fully participate in the following upturn with increasing revenue, profit and valuation. This should allow him to realize the highest possible return.

To determine the current position in the macroeconomic cycle, researchers generally distinguish between early, current and lagged indicators. I consider the first two types in my analysis and neglect the last due to an expected lack of predictive power. A strong current indicator is the industry production growth of a country or region. It is an alternative figure to the GDP growth concerning its predictive value and measures the sum of all goods produced in the industrial sector. Moreover, the business climate, a leading indicator, can be used to proxy the business cycle.

I tested the three variables (GDP growth, industry production growth, business climate); due to multi-collinearity concerns and superior explanatory power for industry production growth, I just show results for this parameter. Other authors have also used GDP growth as a VC performance parameter in their studies. According to Jeng and Wells (2000), who performed a cross-country comparison, there is no significant effect of GDP growth on fund performance. However, Kaplan and Schoar (2005) state that funds raised in boom times usually have no follow-up fund because they often perform poorly. The effect is mainly due to inexperienced fund managers; established funds suffer hardly any performance dilution. Based on these results, I expect a negative effect of industrial production growth on investment performance; investments started during good economic times should perform badly and vice versa, although volatility of returns should be higher in a bad economic environment.

The cycle of an industry which is overlain by the macroeconomic cycle is expected to affect investment returns as well. Quite independent from the general economic development, industries can have technological shocks that lead to superior performance of companies working on these technologies. Other authors (e.g., Buchner et al., 2010) have analyzed the effect of industry specific stock market changes during an investment on the investment performance. They found a positive effect on the average performance of non-defaulting investments. I include the industry specific stock index development[77] before an investment starts in my investigation to proxy for industrial cycles. If the stock index performed very well prior to the investment, I assume that the industry shock has already been priced in. This leads me to an expected negative effect on VC investment returns, increasing default rates and decreasing the average performance of non-defaulting investments.

To evaluate the risk expectations priced in at a certain time, I include the credit spread for general market risks in my investigation. The credit spread is defined as the difference between the investment grade corporate bond rate and the 10-year T-bond rate. Assuming the market acts rationally, a higher risk perception in the market during the time of VC investing should lead to higher shares of defaults and to greater volatility of returns.

To account for further differences in return expectations as well as risks between different industries and investment stages, I include industry and stage dummies in the regressions.

I also tested other variables which measure value creation potential or risk, like the price-earnings-ratio of the S&P 500, at the time of investing, but I excluded them finally due to multi-collinearity concerns.

3.3.4.2 Investment Competition Measures

According to Gompers and Lerner (2000) an increasing availability of capital in the VC market that faces an underproportional growth in (high quality) investment projects leads to price wars and consequently to a decline in investment returns. According to Kaserer and Diller (2007) VC funds are more heavily affected by related market illiquidity than buyout funds.

77 I use Datastream industry indices.

I consider investment competition in my study by analyzing the effect of the (inflation adjusted) total amount of Venture Capital invested in the 12 months before an investment. Contrary to studies by other authors, I do not proxy the competitive intensity with the fundraising volume because even in times with low fund raising volume a high accumulated amount of committed capital may lead to strong investment competition.

During times, when funds invest high amounts of committed capital in investment projects, I expect an increase in default rates, decrease in the average performance and decrease in the chance for exceptional returns. This hypothesis is based on the assumption that higher competition leads investing funds to engage in bidding wars which induce higher prices for portfolio companies and also triggers business models to get financed which are not worth investing under rational considerations.

3.3.4.3 Exit Possibility Measures

After a VC fund manager was able to invest in a portfolio company, foster it to make it attractive for others to buy, the exit market environment is critical for his return. In general the exit via IPO is seen as the most profitable way, more attractive than a trade sale via merger or acquisition. This is confirmed by Peng (2001, p. 33, Table 3). Thus, successful fund managers should plan ahead and anticipate exit conditions just in the moment they invest.

I incorporate expected exit market conditions twofold in my analysis: On the one hand, I consider the absolute amount of VC-backed IPO and M&A deals in the USA in the 12 months prior to the first investment. On the other hand, I assess potential changes in the environment by considering changes to the market conditions 6 months prior to investment. As the IPO and M&A markets both move in cycles the current level and the respective recent change should in combination give a hint in which direction the exit channels are moving, i.e., whether the environment is improving or deteriorating.

I expect that declining/low levels of IPO and M&A activity favor VC investment returns as markets should have recovered by the time of the exit of investments and vice versa.

3.3.4.4 Control Variables

Descriptive analysis reveals that further variables have to be considered to eliminate heterogeneity in the data.

Like other authors (e.g., Nowak et al., 2004 or Giot and Schwienbacher, 2005) I see the time period between Q4/1998 and Q2/2001 as exceptional and related divestment activities as too extreme to generalize. This is why I consider a fixed effect for divestments in this period by including a dummy variable in the regressions.

In order to consider heterogeneity between fund managers in the regression analysis there are generally three options: First, a fixed effect can be included for each General Partner (GP)[78] which increases the degrees of freedom substantially, imbalances estimates and would make predictions impossible; this is not an option. Second, one could model fund manager specifics with random effects.[79] This is not an alternative as GAMLSS does not yet allow random effects methodologically.[80] In consequence, I model heterogeneity between fund managers by considering factors that characterize differences between the fund managers and their funds. Therefore, I include a proxy for the fund managers' experience at the beginning of an investment and the size of the fund in United States Dollar (USD) an investment belongs to in the regressions. Other authors have emphasized the effects of experience and fund size. For example, Gottschalg et al. (2004) identified small fund size to support fund underperformance. At the same

78 Fixed effects can be included in the regressions (equals inclusion of a dummy variable for every fund manager) if one has to conclude that all investments of a fund manager have something in common that is different from investments of other fund managers.

79 Random effect modeling is based on the assumption that the analyzed observations belong of a hierarchy of different populations whose differences are depending on that hierarchy. When investigating systematic differences between fund managers in a regression, inclusion of random effects denotes that each fund manager exerts an individual influence on his investments. The influences/effects of all fund managers together are assumed to follow a given distribution (usually a normal distribution).

80 So far GAMLSS just provides a function that tries to approximate random effects. I do not consider this as reliable enough to use it in my analysis.

time, a too large fund size may lead to opportunistic behavior of the fund managers and consequently to dwindling investment returns. According to Laine and Torstila (2005) the proportion of successful exits is higher for large VC funds compared to small ones. Based on their results I expect a positive influence of experience on investment performance decreasing default rates and increasing the average return as well as the return volatility. Though, two studies provide a rationale for a possible negative effect: Gompers and Lerner (2004) found that reputation concerns make younger partnerships work harder than experienced fund managers to be successful. Schmidt and Wahrenburg (2004) argue that fund managers older and closer to retirement put less effort in their actions to discover and build new business opportunities.

Table 3.1 summarizes the variables used for the analysis including a description of transformations - if applied - to account for non-linearities in the relationship between explanatory and dependent variables.

3.4 Empirical Analysis and Test of the Prediction Methodology

3.4.1 Data Description

3.4.1.1 Composition and Representativeness of Data Sample

The core of my dataset is an extract from the PE deal database of the *Centre for Private Equity Research (CEPRES)*[81]. The CEPRES database was also used by other authors, e.g., Cumming and Walz (2004), Buchner et al. (2010) and Nowak et al. (2004). As of January 2009, it contained information on roughly 23,000 worldwide VC and BO investments from PE funds. The information is completely anonymous, but contains characteristics on investing 'private equity firms', their 'private equity funds' and respective 'portfolio companies'.

81 For further information see www.cepres.de.

Table 3.1: Summary Table of All Variables Used in the Analysis and Prediction of Venture Capital Performance

Category	Variable Name	Description	Transformation used	Source
Performance measures	IRR	Exact Internal Rate of Return based on the monthly investment cashflows	log	CEPRES
	MIRR	Modified Internal Rate of Return based on the monthly investment cashflows assuming a reinvestment/discount rate similar to the annual Nasdaq performance between 1985 and 2003	log	own calculation
GP level	FundNumber	The sequence number of the general partners' fund an investment belongs to	log	CEPRES
Fund level	FundSize	Amount of committed capital of fund an investment belongs to (in USD)	log	CEPRES
Portfolio Company level	EarlyStage	A dummy variable equal to 1 for early stage investments (Companies belonging to the CEPRES stage categories "Seed", "Start-up" and "Early"), 0 otherwise (Early stage is the reference category)	dummy	CEPRES
	LaterStage	A dummy variable equal to 1 for later stage investments (Companies belonging to the CEPRES stage categories "Later" or "Expansion"), 0 otherwise	dummy	CEPRES
	UnspecifiedVC	A dummy variable equal to 1 for unspecified VC investments (Companies belonging to the CEPRES stage category "Unspecified VC"), 0 otherwise	dummy	CEPRES
	ShareEarlyStage	Share of early stage investments within all investments of a quarter (Early stage investments include companies belonging to the CEPRES stage categories "Seed", "Start-up" and "Early")	dummy	CEPRES
	Biotechnology	A dummy variable equal to 1 for companies of the Biotechnology-Sector (for classification see Table 13 in the appendix), 0 otherwise (Biotechnology is the reference category)	dummy	CEPRES
	Industrials	A dummy variable equal to 1 for companies of the Industrials-Sector, 0 otherwise	dummy	CEPRES
	Computer	A dummy variable equal to 1 for companies of the Computer-Sector, 0 otherwise	dummy	CEPRES
	Software	A dummy variable equal to 1 for companies of the Software-Sector, 0 otherwise	dummy	CEPRES
	Internet&Media	A dummy variable equal to 1 for companies of the Internet&Media-Sector, 0 otherwise	dummy	CEPRES
	Telecom	A dummy variable equal to 1 for companies of the Telecommunication-Sector, 0 otherwise	dummy	CEPRES
	Other	A dummy variable equal to 1 for companies of other or unspecified sectors, 0 otherwise	dummy	CEPRES

Table 3.1: Summary Table of All Variables Used in the Analysis and Prediction of Venture Capital Performance (cont'd)

Level	Variable Name	Description	Transformation used	Source
Macro-economic	IndustrialProduction	Change of the United States of America (USA) industrial production in the 12 months before the first investment	identity	Datastream
	IndustryPerformance	Discrete annual return (in %) of the Datastream Industry Index matching the CEPRES industry classification (see Table 13 in the appendix) in the 12 months before the first investment	log	Datastream
	MeanIndustryPerformance	Mean of the discrete annual return (in %) of the Datastream Industry Index matching the CEPRES industry classification across the investments of a quarter (see Table 13 in the appendix) in the 12 months before the first investment	identity	Datastream
	CreditSpread	Average spread of investment grade US corporate yield over 10-year treasury bond yield in 12 months before the first investment	identity	Datastream
	VCInvestmentVolume	Total volume of VC investments in the USA in the 12 months before the first investment (figure is inflation (CPI) adjusted to 1983 USD)	log	VentureXpert
	IPOs	Number of (PE-backed) IPOs in the USA in the 12 months before the first investment	identity	VentureXpert
	ChangeIPOs	Change in the number of (PE-backed) IPOs in the USA in the 12-7 months to the 6-0 months before the first investment	identity	VentureXpert
	ChangeM&A	Change in the number of M&A deals in the USA in the 12-7 months to the 6-0 months before the first investment	identity	Thompson One
	DivestBubble	A dummy variable equal to 1 for investments finally exited between September 1998 and March 2000, 0 otherwise	dummy	CEPRES

For the current investigation, I just include Venture Capital investments of worldwide PE fund managers in companies in the United States of America who made their first investments into a portfolio company in the period between 1985 and 2003. The USA are an especially interesting area for investigations on VC: On the one hand, it has the longest history in (modern) VC investing. On the other hand, it is a fairly homogeneous economic area (especially more homogeneous than Europe with its different legal environments).[82] The time span between 1985 and 2003 was chosen because sufficient information on the economic environment as well as a sufficient number of investments in the CEPRES database per year were not available before 1985. Investments after 2003 were excluded due to a possible survivorship bias. The mean holding period of a realized investment in my data sample was 46 months. Consequently, investments started in 2004 and thereafter have a high chance that fund managers didn't exit the investments until the beginning of this analysis in 2009.[83] I explicitly included the 1999/2000 Internet hype into the dataset to investigate whether or not the irregularities in this time span can be partially explained and modeled by common patterns.

The analysis of VC investments relies on industry insiders to reveal their private transactions. Thus, it is generally unknown whether an investigated data sample is representative for the whole industry or biased. There are two basic approaches to overcome this issue: Either one can - at least partially - statistically control for the selection bias on small data samples, e.g., by incorporating a Heckman model[84], or enlarge the data sample and check its' representativeness using different indicators. Because I face the problem of

82 For a comparison between European and North American VC investments please refer to Kaserer and Diller (2004) and Hege et al. (2009).

83 The study was performed in Q2/2009 which is 5.5 years after the start of the investment. Then, an average VC investment is already sold or written-off. Market conditions up to 2006 were very favorable for successful VC exits and badly performing investments should already have been exited after 2 years of market downturn.

84 Other authors, e.g., Cumming and Walz (2004), try to account for selection bias by applying a Heckman methodology. Usually they regard the Thompson VentureXpert database as the most comprehensive source on VC funds, and rebalance their recordsets in the analysis by overweighing underrepresented groups of recordsets and vice versa.

anonymity of my data sample, I cannot apply the Heckman approach. My advantage is that my sample is large and thus I check the representativeness descriptively by comparing it with the Thompson VentureXpert database.

Table 3.2 shows the distribution of the GPs' founding year (Panel A), vintage years[85] of the funds (Panel B) and investment years of the portfolio companies by stages and industries (Panel C). Keeping in mind that many new GPs started their business in the late 1990's during the technology and Internet boom, Panel A confirms that the data sample is quite balanced concerning older and younger GPs. In total 64 partnerships in the sample were established before 1995 and 51 after 1995.

As the timely distribution of the funds' vintage years indicates, the data sample is quite similar to the distribution of vintage years in the VentureXpert database. The vintage years at the end of the 1980's are a little underrepresented in the data as are the vintage years 2002 and 2003. The later is not surprising as funds from these vintage years which didn't invest in 2002/2003 and fully exit their investments until 2009 do not appear in the sample. These observations indicate that the probability of a skewness through over- or under-representativeness of funds of certain vintage years seems low. Though, a bias towards larger funds might exist in the dataset. Comparing the total fund size in my sample with the total fund size in the VentureXpert continuum indicates that 27% of the VentureXpert's fund size is in my sample. However, I just have 6% of the VentureXpert's number of funds in my sample.

Concerning investment stage, I divide the VC dataset into Early Stage (seed and start-up stages) and Later stage (incl. expansion stage). According to Panel C, the majority of the 3,479 investigated investments (53%) belongs to the Early Stage group, 29% are in the Later Stage group and for 17% of investments the classification is unknown. There is quite little variation in these shares over time.

The classification into industries is based on the CEPRES database classification. As the number of industries in the database is too large to yield

85 The vintage year is the year a VC fund has its' first investment.

Table 3.2: Structure of the Dataset

Table shows the distribution of the investing General Partners' founding years (Panel A), their funds' vintage years (Panel B) and respective Venture Capital investments by stage, country and industry focus (Panel C).

Panel A: Founding Years of General Partnerships

Founding Year	1960-1969	1970-1979	1980-1984	1985-1989	1990-1994	1995	1996	1997	1998	1999	2000-2003
No. of GPs	4	10	17	18	15	5	9	8	8	10	11

Panel B: Vintage Years of Sample Funds in comparison to VentureXpert

Vintage Year	1981-1984	1985	1986	1987	1988	1989	1990	1991	1992	1993	1994	1995	1996	1997	1998	1999	2000	2001	2002	2003	Total
CEPRES Sample																					
No. of VC funds	8	4	2	3	5	7	6	5	7	6	9	15	14	23	32	40	53	13	5	1	258
% of VentureXpert sample	*2%*	*3%*	*2%*	*2%*	*5%*	*7%*	*6%*	*12%*	*8%*	*6%*	*6%*	*8%*	*8%*	*8%*	*10%*	*8%*	*7%*	*3%*	*2%*	*0%*	*6%*
Total size VC fund ($ Mio)	868	102	1,281	213	400	2,218	316	325	551	1,101	2,544	2,308	1,424	2,785	11,329	18,630	18,200	6,532	1,721	40	72,886
% of VentureXpert sample	*8%*	*3%*	*34%*	*5%*	*9%*	*42%*	*9%*	*16%*	*10%*	*25%*	*28%*	*23%*	*12%*	*14%*	*37%*	*33%*	*17%*	*16%*	*16%*	*0%*	*20%*
No. of Buyout funds	0	0	0	2	1	2	0	1	1	3	0	2	6	3	3	4	4	2	1	0	35
Total size Buyout fund ($ Mio)	0	0	0	236	92	794	0	133	314	564	0	1,378	2,003	2,142	3,897	1,425	6,865	2,352	325	0	22,520
VentureXpert sample (VC in North America)																					
No. of funds	440	124	104	121	107	107	95	42	87	98	144	189	181	287	334	522	758	391	252	202	4,585
Total size ($ Mio)	10,392	3,980	3,791	4,611	4,524	5,283	3,532	2,003	5,504	4,385	9,149	10,043	12,000	20,170	30,269	56,967	106,645	39,913	10,528	11,952	355,639

Panel C: Distribution of Sample Investments

Investment Year	1985	1986	1987	1988	1989	1990	1991	1992	1993	1994	1995	1996	1997	1998	1999	2000	2001	2002	2003	Total
by Stage																				
Early stage	19	29	42	50	65	33	37	54	69	79	86	103	134	184	296	314	109	83	67	1853
% of total	*37%*	*43%*	*61%*	*47%*	*50%*	*35%*	*42%*	*47%*	*53%*	*64%*	*54%*	*51%*	*52%*	*54%*	*53%*	*55%*	*61%*	*64%*	*68%*	*53%*
Later stage	20	24	17	37	48	45	38	37	37	27	47	68	86	103	156	140	39	28	21	1018
% of total	*39%*	*36%*	*25%*	*35%*	*37%*	*48%*	*43%*	*32%*	*28%*	*22%*	*30%*	*34%*	*33%*	*30%*	*28%*	*24%*	*22%*	*22%*	*21%*	*29%*
Unspecified VC	12	14	10	19	16	16	14	24	25	17	26	30	37	56	110	122	32	18	10	608
% of total	*24%*	*21%*	*14%*	*18%*	*12%*	*17%*	*16%*	*21%*	*19%*	*14%*	*16%*	*15%*	*14%*	*16%*	*20%*	*21%*	*18%*	*14%*	*10%*	*17%*
by Sector																				
Computer	14	22	13	14	21	10	9	16	15	21	28	25	53	56	103	145	57	43	31	696
% of total	*27%*	*33%*	*19%*	*13%*	*16%*	*11%*	*10%*	*14%*	*11%*	*17%*	*18%*	*12%*	*21%*	*16%*	*18%*	*25%*	*32%*	*33%*	*32%*	*20%*
Software	5	5	5	10	12	13	19	15	21	23	19	66	59	82	119	108	47	39	28	695
% of total	*10%*	*7%*	*7%*	*9%*	*9%*	*14%*	*21%*	*13%*	*16%*	*19%*	*12%*	*33%*	*23%*	*24%*	*21%*	*19%*	*26%*	*30%*	*29%*	*20%*
Internet and Media	2	7	6	6	1	4	1	4	5	7	14	15	22	43	145	104	10	5	7	408
% of total	*4%*	*10%*	*9%*	*6%*	*1%*	*4%*	*1%*	*3%*	*4%*	*6%*	*9%*	*7%*	*9%*	*13%*	*26%*	*18%*	*6%*	*4%*	*7%*	*12%*
Telecom	13	6	9	11	26	4	2	23	16	20	27	36	40	53	78	136	34	20	6	560
% of total	*25%*	*9%*	*13%*	*10%*	*20%*	*4%*	*2%*	*20%*	*12%*	*16%*	*17%*	*18%*	*16%*	*15%*	*14%*	*24%*	*19%*	*16%*	*6%*	*16%*
Biotechnology	9	11	18	24	26	34	26	32	38	31	40	29	54	68	59	42	15	10	9	575
% of total	*18%*	*16%*	*26%*	*23%*	*20%*	*36%*	*29%*	*28%*	*29%*	*25%*	*25%*	*14%*	*21%*	*20%*	*10%*	*7%*	*8%*	*8%*	*9%*	*17%*
Industrials	5	13	16	26	28	23	18	18	17	15	20	22	12	12	14	12	6	3	4	284
% of total	*10%*	*19%*	*23%*	*25%*	*22%*	*24%*	*20%*	*16%*	*13%*	*12%*	*13%*	*11%*	*5%*	*3%*	*2%*	*2%*	*3%*	*2%*	*4%*	*8%*
Other and Unspecified	3	3	2	15	15	6	14	7	19	6	11	8	17	29	44	29	11	9	13	261
% of total	*6%*	*4%*	*3%*	*14%*	*12%*	*6%*	*16%*	*6%*	*15%*	*5%*	*7%*	*4%*	*7%*	*8%*	*8%*	*5%*	*6%*	*7%*	*13%*	*8%*
Total	**51**	**67**	**69**	**106**	**129**	**94**	**89**	**115**	**131**	**123**	**159**	**201**	**257**	**343**	**562**	**576**	**180**	**129**	**98**	**3479**
% of all investments	*1%*	*2%*	*2%*	*3%*	*4%*	*3%*	*3%*	*3%*	*4%*	*4%*	*5%*	*6%*	*7%*	*10%*	*16%*	*17%*	*5%*	*4%*	*3%*	*100%*

useful results in my analysis, I combine the original industry codes to the seven sectors Biotechnology, Telecom, Industrials, Computer, Software, Internet & Media and Other & Unspecified.[86] The development of the split of investments into sectors over time is as expected. Software and Internet & Media investments gained significantly in the end of the 1990's while the number of Biotechnology and Industrials investments that dominated in the beginning of the 1990's faded over time.

The aggregate distribution of the investments over time is in line with what one would expect: During the late 1980's and the early 1990's the VC industry grew significantly in the USA. In the end of the 1990's and especially 1999 and 2000 was the time of most VC activity followed by a sharp decline in the new millennium. This development is reflected in the number of investments in the dataset; about a third of investments was started in 1999/2000 with a general growth trend between 1985 and 2000. Figure 3.3 shows that in my dataset the distribution of investments over time is very similar to the distribution of first-round PE investments in the VentureXpert database which is considered to be quite comprehensive. Just in the years 1991-1994 and 2002/2003 my data sample seems to be a little underrepresented.

3.4.1.2 Distribution Pattern of Investment Performance

Table 3.3 shows the distribution pattern of direct investment performance measured in IRR (Panel A), MIRR (Panel C) and respective logarithms (Panels B and D). Of the 3,479 investigated investments 32% were written off completely. The IRR distribution pattern of investments which didn't default, ranges from -99% to 1,481% annual IRR.[87] The median IRR of the sample is 10.8%, positive, but far below the mean of 51.8%. This effect is a result of the top 10% of the IRRs which are greater than 100%. The facts

86 Other authors, e.g., Buchner et al. (2010), have classified the CEPRES industries into five sectors; my approach differs as I separate the high-tech sectors computer, software, Internet & media due to differences in industry cyclicality and involved risks. Hardware business involves totally different economics than software business. Internet & Media both relying heavily on revenues from advertising are expected to display different chance-risk-relationships.

87 There were a few data points in the CEPRES data of IRRs above 1,500%. I excluded them in the regression analysis as they would imbalance regression estimates too heavily.

Figure 3.3: Comparison of the Distribution of Investments over Time between the CEPRES Dataset and the VentureXpert Database
The blue line includes all worldwide first-round PE investments in the time from 1985 to 2003 listed in the VentureXpert database. The red line includes all investments used for this study.

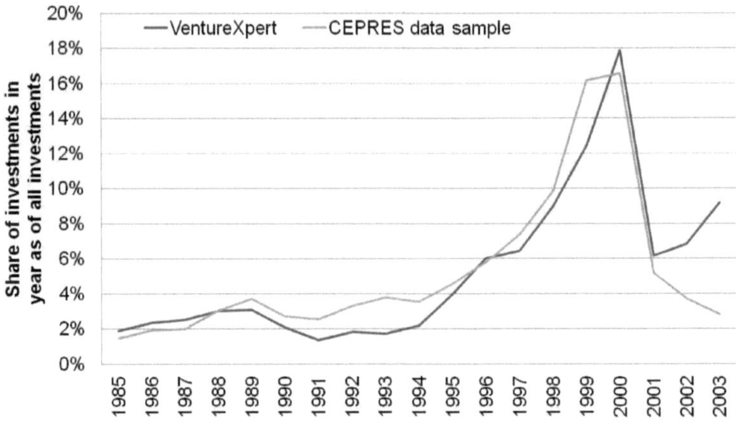

that the 75 percentile is close to the mean and the high standard deviation of 182% underline the high positive skewness (at 4.02) of the IRR distribution pattern. The kurtosis value of 20.38 indicates a high leptokurtosis, i.e., thin but long tails.

The general form of the IRR distribution pattern is similar across investment stage and sector clusters; differences exist in the extent: As one would expect, early stage investments have a higher default rate (36%) than later stage investments (23%). The high default rate (35%) of investments with unspecified investment stage points to a high share of early stage investment in this group. Both the median and the mean of early stage investment IRRs are below the values of later stage investments, indicating less attractive investment opportunities – especially when one considers the higher default risks of early stage investments.

Fundamental differences in business models of the portfolio companies – indicated by the sector they belong to, respectively – leads to significant differences in default rates, skewness and kurtosis. Biotechnology investments have the lowest share of write-offs (23%), while more than a third of Software, Internet & Media as well as Telecom investments had to be written off completely. The groups with the lowest median return of non-defaults were Internet & Media and Computer investments. For Internet & Media this is even negative at -3.4%. The weakest mean performance of non-defaults is indicated for investments in the Industrials cluster at 27.9%; the comparatively low standard deviation of 90% indicates that hardly any "super" returns were generated in this cluster.

Panel B shows the results of a logarithmic transformation of the IRR figures. It is interesting to note that after the transformation, the mean of 4.61 of all investments equals a return of 0% ($log(100) = 4.6052$). Despite the transformation the performance distribution pattern is still not close to normality. The skewness turns negative and the kurtosis is still significantly above zero.[88]

When the performance of investments is measured with MIRR the skewness and kurtosis is even greater than with IRR across investment stage and sector clusters (see Panel C). Due to the different approach concerning the reinvestment assumption, the maximum MIRRs are below the maximum IRRs. Furthermore, mean performances of non-defaults decrease across clusters.

As with IRRs, a logarithmic transformation of MIRR figures shifts the distribution pattern closer to normality, but still a significant skewness and kurtosis remain in the data (see Panel D).

How VC investments of different investment years performed historically between 1985 and 2003 is displayed in Figure 3.4.[89] The top panel shows the distribution pattern with a box plot per year. The chart on the bottom illustrates the share of investments that fall into specific IRR ranges. The red area shows the development of defaults, the red and yellow areas together are the share of negative returns. Any area above the green part represent investments that realized higher IRRs than 100%. In the time between 1985 and 1991, the return distribution patterns were quite stable, but then the overall volatility constantly increased. This led to a higher

88 The kurtosis in this table is standardized to zero for a normal distributed figure.

89 The distribution pattern of returns measured with MIRR is not shown here but fairly close to the IRR distribution pattern.

Table 3.3: Distribution Patterns of Return Measures IRR, log(IRR), MIRR and log(MIRR)

Panel A: IRR (in %)

	N	% of total N	% Write-Offs	Non-Write-Offs								
				Min	25%	Mean	Median	75%	Max	Std	Skewness	Kurtosis
All Investments	3,479	100%	32%	-99.0	-32.7	51.8	10.8	54.4	1,481.5	182.4	4.02	20.38
Stage Cluster												
Early Stage	1,853	53%	36%	-99.0	-39.7	49.3	2.6	50.4	1,481.5	191.3	3.94	18.88
Later Stage	1,018	29%	23%	-99.0	-12.4	58.3	19.0	62.6	1,481.3	169.7	4.30	24.98
Unspecified	608	17%	35%	-97.3	-45.5	46.4	10.3	46.9	1,288.0	179.5	3.78	17.35
Sector Cluster												
Biotechnology	575	17%	23%	-96.0	-13.9	42.3	15.6	54.1	1,249.9	146.4	5.29	34.45
Industrials	284	8%	29%	-91.7	-6.0	27.9	18.0	41.7	989.5	91.0	6.16	60.01
Computer	696	20%	33%	-98.0	-43.0	57.6	4.8	59.3	1,481.5	196.0	3.39	14.60
Software	695	20%	35%	-99.0	-29.1	53.6	10.8	60.1	1,481.3	180.9	3.88	19.81
Internet & Media	408	12%	39%	-99.0	-55.8	43.5	-3.4	59.4	1,102.0	167.3	2.66	9.08
Telecom	560	16%	34%	-98.9	-38.4	82.7	13.5	62.3	1,478.2	257.9	3.16	10.86
Other & Unspecified	261	8%	30%	-93.3	-22.4	30.0	13.0	43.0	1,190.8	121.7	5.93	48.55

Panel B: log(100 + IRR (in %))

	N	% of total N	% Write-Offs	Non-Write-Offs								
				Min	25%	Mean	Median	75%	Max	Std	Skewness	Kurtosis
All Investments	3,479	100%	32%	0.70	4.22	4.61	4.72	5.05	7.37	0.94	-0.56	1.92
Stage Cluster												
Early Stage	1,853	53%	36%	0.71	4.12	4.55	4.64	5.02	7.37	0.99	-0.54	1.83
Later Stage	1,018	29%	23%	0.70	4.48	4.74	4.79	5.10	7.37	0.85	-0.68	2.63
Unspecified	608	17%	35%	1.30	4.02	4.56	4.71	5.00	7.24	0.95	-0.28	0.94
Sector Cluster												
Biotechnology	575	17%	23%	1.60	4.47	4.71	4.76	5.04	7.21	0.71	-0.45	3.26
Industrials	284	8%	29%	2.23	4.55	4.69	4.78	4.96	6.99	0.63	-1.30	4.14
Computer	696	20%	33%	1.10	4.06	4.57	4.66	5.08	7.37	1.02	-0.30	0.72
Software	695	20%	35%	0.70	4.28	4.60	4.72	5.08	7.37	1.00	-0.79	2.10
Internet & Media	408	12%	39%	0.71	3.81	4.45	4.58	5.08	7.09	1.09	-0.39	0.44
Telecom	560	16%	34%	0.75	4.14	4.61	4.74	5.10	7.36	1.13	-0.35	1.16
Other & Unspecified	261	8%	30%	2.04	4.36	4.65	4.74	4.97	7.16	0.69	-0.60	2.79

Table 3.3: Distribution Patterns of Return Measures IRR, log(IRR), MIRR and log(MIRR) (cont'd)

Panel C: MIRR (in %)	N	% of total N	% Write-Offs	Non-Write-Offs								
				Min	25%	Mean	Median	75%	Max	Std	Skewness	Kurtosis
All Investments	3,479	100%	32%	-98.9	-21.7	29.0	11.1	40.4	1,154.5	114.7	4.88	34.23
Stage Cluster												
Early Stage	1,853	53%	36%	-98.9	-24.8	24.4	5.4	38.2	1,116.8	111.6	5.03	36.89
Later Stage	1,018	29%	23%	-97.8	-6.1	37.3	17.3	46.1	1,140.2	112.7	4.51	29.44
Unspecified	608	17%	35%	-95.1	-35.4	26.3	10.6	35.7	1,154.5	126.6	5.05	34.65
Sector Cluster												
Biotechnology	575	17%	23%	-90.2	-7.3	30.7	14.8	40.2	1,154.5	111.9	6.51	53.96
Industrials	284	8%	29%	-89.3	0.1	23.5	16.8	33.5	984.9	83.2	7.87	86.58
Computer	696	20%	33%	-98.0	-29.3	24.0	6.7	42.3	781.3	101.1	3.39	17.17
Software	695	20%	35%	-98.2	-19.3	31.1	11.1	44.6	1,116.8	118.9	4.80	32.63
Internet & Media	408	12%	39%	-95.6	-44.0	28.4	0.0	42.1	903.5	122.2	2.90	12.57
Telecom	560	16%	34%	-98.9	-26.9	40.7	13.0	46.1	1,140.2	154.1	4.10	21.37
Other & Unspecified	261	8%	30%	-79.1	-13.6	15.8	13.0	31.2	267.9	51.1	1.33	3.88

Panel D: log(100 + MIRR (in %))	N	% of total N	% Write-Offs	Non-Write-Offs								
				Min	25%	Mean	Median	75%	Max	Std	Skewness	Kurtosis
All Investments	3,479	100%	32%	0.75	4.37	4.63	4.72	4.95	7.14	0.70	-0.74	3.48
Stage Cluster												
Early Stage	1,853	53%	36%	0.75	4.33	4.60	4.67	4.94	7.10	0.71	-0.81	3.95
Later Stage	1,018	29%	23%	1.15	4.55	4.72	4.77	4.99	7.12	0.66	-0.79	3.92
Unspecified	608	17%	35%	1.78	4.18	4.57	4.71	4.92	7.14	0.75	-0.40	1.77
Sector Cluster												
Biotechnology	575	17%	23%	2.38	4.54	4.72	4.75	4.95	7.14	0.53	0.08	4.23
Industrials	284	8%	29%	2.46	4.62	4.70	4.77	4.90	6.99	0.50	-0.97	5.34
Computer	696	20%	33%	1.11	4.27	4.59	4.68	4.97	6.78	0.73	-0.68	2.15
Software	695	20%	35%	1.04	4.40	4.64	4.72	4.98	7.10	0.73	-0.87	3.57
Internet & Media	408	12%	39%	1.69	4.04	4.52	4.62	4.96	6.91	0.87	-0.50	1.17
Telecom	560	16%	34%	0.75	4.31	4.61	4.74	4.99	7.12	0.85	-0.63	2.72
Other & Unspecified	261	8%	30%	3.09	4.47	4.67	4.74	4.88	5.91	0.46	-0.62	1.07

share of write-offs and temporarily to exceptional high returns of more than 1,000% p.a. for investments started in 1997-1999, i.e., during the technology bubble, followed by default rates of about 50% for investments started in 2000. In the aftermath of the bubble the default rates decreased slightly but volatility remained high. One could argue that the increased volatility is an effect of the higher amount of observations in the data sample for the later time period, but my statistical analysis in the following sections will show the contrary. Differences in volatility may be explained by changes in the investment environment.

3.4.2 Parameters of Investment Performance

3.4.2.1 Default Probability Estimated in Analogy to Credit Portfolio View

As explained in the methodology section, the probability that a single investment has to be written off completely may be estimated with a LOGIT regression model. Table 3.4 shows the results of three model specifications with a dummy variable which indicates if an investment was written off completely (dummy=1) or not (dummy=0) as the variable to be explained. Due to multi-collinearity concerns between the variables CreditSpread, IPOs and VCInvestmentVolume I tested different specifications – one including CreditSpread and IPOs and two including VCInvestmentVolume. As my goal is to get a good predictive power of my model, I especially investigated the functional form of the influence of the independent variables. I went beyond the widespread approach of taking the logarithm of independent variables to get a more linear relation, but partially fitted orthogonal polynomial splines and checked the stability of the functional forms across multiple specifications (for more details on robustness see Section 3.4.2.4). Figure 3.5 shows the partial effects of the covariates in Specification 3 of Table 3.4 including standard errors (indicated with dotted lines). Specification 3 is used throughout the rest of this study to test the power of the

Figure 3.4: Distribution Pattern of Performance (IRR) of Venture Capital Investments between 1985 and 2003

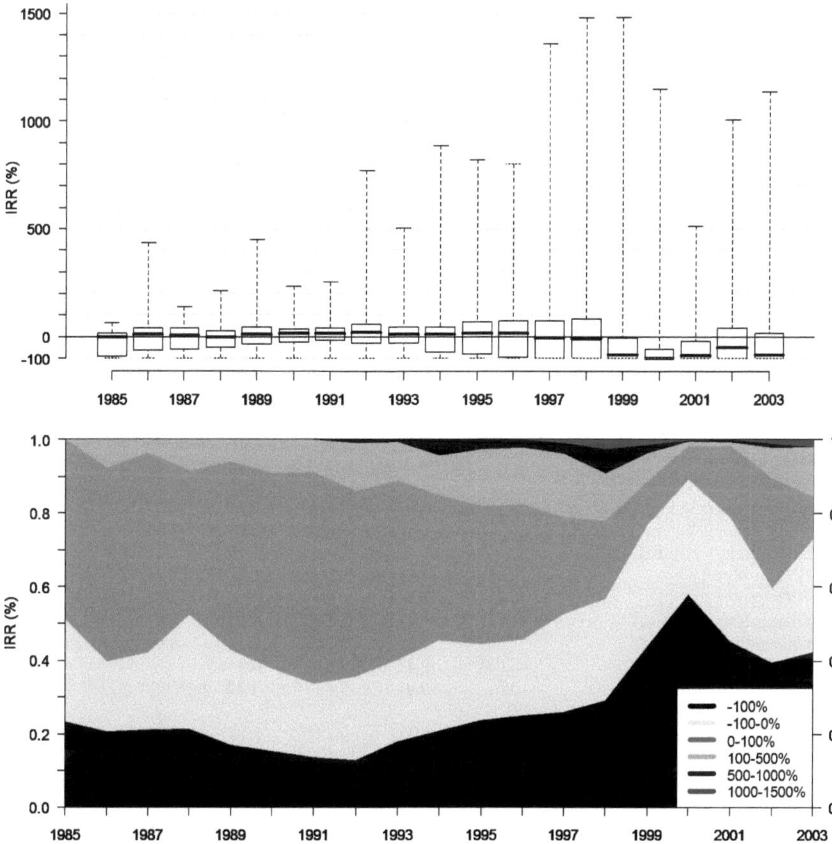

Table 3.4: Parameters of the Default Rate of Individual Investments (Credit Portfolio View)

The table shows the result of three specifications of LOGIT regressions for a dataset of US VC investments of PE funds between 1985 and 2003. Dependent variable is a dummy variable that is 1 if an investment was written off completely, i.e. IRR=-100%, and 0 otherwise. Observations are weighted with the relative share of capital which is invested. Independent variables are macroeconomic parameters of performance indicating investment chances, risks, investment competition and exit market conditions at the time of the first investment into the portfolio company. To control for dependencies from specifics of the investing fund manager and fund, control variables FundSize and FundNumber are included in the specifications. Variables are as explained in Table 3.1. Some covariates are fitted with orthogonal polynomials which is indicated by "(poly=Z)"; Z stands for the orthogonal polynomial of the Zth degree. The indicators *, **, *** signal significance at the 10%, 5% and 1% levels, respectively.

	Dependent variable: Write-Off-Dummy (IRR=-100%)					
	Specification 1		**Specification 2**		**Specification 3**	
	Coefficient	z-value	Coefficient	z-value	Coefficient	z-value
Chances						
IndustrialProduction (poly=1)	26.10***	6.51	8.66***	2.60	9.44***	3.43
IndustrialProduction (poly=2)	0.17	0.06	-7.33***	-2.84	-7.58***	-3.10
log(IndustryPerformance)	-0.06	-0.52	-0.04	-0.35		
Risks						
CreditSpread (poly=1)	24.82***	6.71				
CreditSpread (poly=2)	-4.63	-1.51				
CreditSpread (poly=3)	-12.03***	-4.29				
LaterStage	-0.52***	-5.93	-0.49***	-5.55	-0.48***	-5.50
UnspecifiedVC	0.03	0.24	-0.04	-0.30	-0.03	-0.26
Industrials	0.72***	3.50	0.90***	4.26	0.89***	4.21
Computer	0.62***	4.09	0.47***	3.06	0.45***	2.96
Software	0.73***	4.70	0.58***	3.70	0.57***	3.65
Internet&Media	0.41**	2.33	0.29*	1.67	0.24	1.48
Telecom	0.56***	3.71	0.38**	2.49	0.36**	2.37
Other	0.36*	1.88	0.31	1.57	0.30	1.52
Competition						
VCInvestmentVolume (poly=1)			43.76***	12.22	42.77***	12.26
VCInvestmentVolume (poly=2)			-7.66**	-2.53	-6.37**	-2.16
VCInvestmentVolume (poly=3)			-9.58***	-2.61	-7.66**	-2.24
Exit market conditions						
IPOs (poly=1)	2.68	0.93				
IPOs (poly=2)	6.42**	2.24				
ChangeIPOs (poly=1)	11.11***	3.95	-7.15**	-2.21	-5.21*	-1.75
ChangeIPOs (poly=2)	7.68***	2.74	5.81**	2.11	6.00**	2.20
ChangeM&A (poly=1)	-6.63**	-2.42	2.67	0.94		
ChangeM&A (poly=2)	-0.01	-0.01	-2.89	-1.12		
Control variables						
DivestBubble	0.53***	3.90	0.72***	5.13	0.70***	5.03
FundSize (poly=1)	-4.81	-1.17	-11.07***	-2.62	-10.99***	-2.61
FundSize (poly=2)	10.56***	3.60	13.38***	4.46	13.46***	4.51
FundNumber (poly=1)	4.74*	1.82	4.81*	1.85	4.88*	1.88
FundNumber (poly=2)	1.21	0.51	-0.99	-0.41	-1.00	-0.42
FundNumber (poly=3)	9.40***	-3.97	-8.78***	-3.69	-8.58***	-3.61
(Intercept)	-0.95	-1.61	-1.04*	-1.82	-1.22***	-9.29
Model diagnostics						
Observations	3,479		3,479		3,479	
AIC	4,000		3,922		3,920	
Chi-Square	318		396		393	
p-value	0		0		0	
Pseudo R^2	12.10%		14.90%		14.80%	

model; it seems a fair balance between the number of covariates and the explanatory power (R^2 is at 14.8%) indicated by a comparatively low AIC.

The partial effects indicate a generally positive significant effect of the industrial growth on the probability of default which is in line with my expectations. The more the economy is roaring at the moment of the first investment into a portfolio company, the higher the expectation that the company cannot be sold successfully afterwards. The relationship has a diminishing effect. The increase in defaults is stronger when the economy goes from bad to good than when it goes from good to very good. This positive effect of the industrial growth may be explained by economic cycles. If a fund manager invests in a company during a period of strong economic growth, the growth will probably diminish in the following period. Thus, the chances to successfully build the business of the venture is lower which increases the default probability.

The negative influence of the industry performance in the 12 months prior to the investment on the write-off probability is surprising, but the influence is not statistically significant in any specification.

However, the effect of the investment risk profile on the default probability is highly significant. Early stage investments (which are the control group) have a higher default rate than investments in later stage businesses. Different industries also seem to have different risks in venture investing. The control group "Biotechnology" has the lowest risk, "Software", "Industrials" and "Computer" the highest. These results are in line with the descriptive analysis.

A highly significant effect is revealed for the credit spread. Higher general risk in the market leads to an increasing share of defaults. Though, this effect is diminishing for very high credit spreads. At a first glance this might surprise as an increasing risk in the market should lead to more defaults. A possible explanation of my result is, that once the risk premium passes a certain threshold (indicated at around 280 basis points), the fund managers' increased risk aversion leads to more restrictive investment behavior.

Figure 3.5: Functional Form of Partial Effects of Covariates in Specification 3 of Table 3.4

The vertical axis of the charts plots the partial effects of the covariates. Dotted lines indicate standard errors.

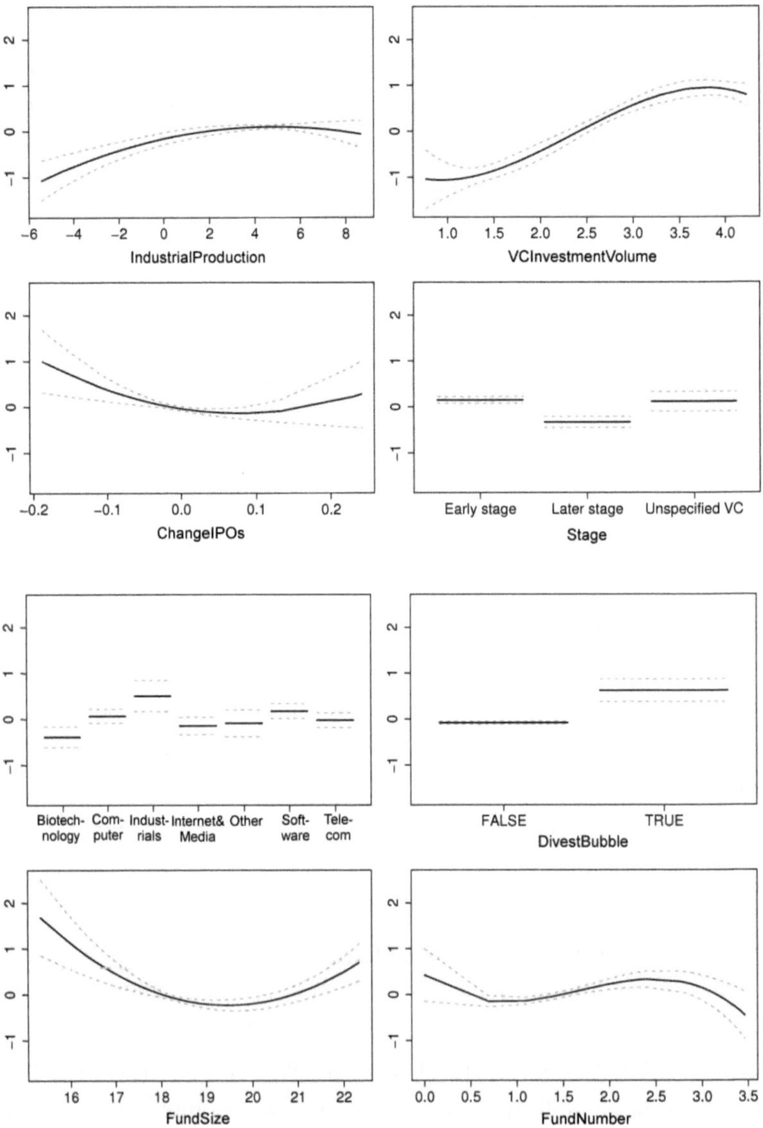

They just spend capital on ventures that seem a save bet. This hypothesis is supported by the analysis of the non-defaults (see Section 3.4.2.3) which shows that an increasing credit spread leads to a reduction in the dispersion of IRRs.

The variable approximating bidding wars – VCInvestmentVolume – is highly significant as well. The not tabulated analysis of variance (ANOVA) indicates that this variable has the highest share of explanatory power. The relationship between the logarithm of the investment volume in the 12 months before the first investment in a portfolio company and the default probability is almost linear increasing in line with my expectations. The more money is chasing the few available (favorable) deals in the market, the greater the fraction of worse deals that also get financing and the more companies are liquidated instead of sold to other investors. For very high investment volumes, the effect is also slightly diminishing as with the high credit spread. As both variables CreditSpread and VCInvestmentVolume have a quite high correlation one may interpret this effect as an adaptation of investment policy as well.

The constitution of the exit markets, IPO and M&A, at the time of the first investment in a portfolio company are strong indicators for defaults. If the IPO market is very strong or heavily increasing, the probability of default rises with the market constitution as I expected. Weakening IPO market conditions also lead to increasing defaults, i.e., a moderately growing number of IPOs is most favorable. This result probably originates from the IPO cycles which coincide with economic cycles. If fund managers invest in "hot" times, the investment will more probably be a failure. The functional form of the effect of "ChangeM&A" is almost linearly decreasing though not statistically significant in Specification 2. M&A markets are generally less cyclical and more trending as indicated by my data. A positive market development holds for a longer period of time as does a negative development. Thus, an improving M&A environment at the time of first investing in a venture seems to hold long enough to give fund managers a chance for a sale - however, this does not necessarily imply that a high return may be generated.

The dummy variable DivestBubble which indicates a sale during the Internet bubble is positive and highly significant. This might surprise on the first sight as one would expect that almost anything could be sold during the bubble reducing the share of defaults. The contrary is the case. Especially during these hot market conditions, fund managers seemed to have focused their attention on "star" investments that had the chance to generate exceptionally high returns and wrote off "living deads" very quickly. This result supports the signalling hypothesis. During good economic times, when fund managers can impress their investors with great returns of some investments, it is easier for them to argue misinvestments.

Concerning the control variable FundSize a convex relationship is indicated with an optimum at around USD 300M ($e^{19.5}$), i.e., bigger and smaller funds generate more defaults. This result is in line with findings from Kaplan and Schoar (2005, p. 1804) on the fund level. For smaller funds there are economies of scale. In bigger funds the GPs may not be able to sufficiently support investments any more - which is important for young ventures. The number of highly skilled managers underproportionally increases with the number of managed investments.

My data unveil a significant but surprising influence of experience on defaults. First-time funds and medium experienced fund manager (with about their 13th fund) produce the highest shares of write-offs. Funds management teams with their 3rd fund and GPs with long-term experience (e.g., with their 30th or more fund) experience rather low defaults. This could be interpreted as a result of a learning process. Inexperienced fund managers seem to take too much risk. After burning their fingers in their 1st and 2nd fund they become restrictive in their behavior and take less risk. After GPs have (successfully) invested over a longer period of time and their funds have grown in size, GPs again may take more risks to get their funds invested or they change their write-off policy and cancel unsuccessful investments earlier. Very experienced GPs with usually big funds and corresponding big organizations need to build more bureaucratic structures which may induce more aversion against risky investments. (Although this seems a feasible explanation I cannot present a further proof due to data limitations.)

3.4.2.2 Default Probability Estimated in Analogy to the Credit Risk Evaluation Model

An alternative to the modeling of individual defaults with the CPV based approach is to model defaults with the Credit Risk Evaluation Model (CREM) based approach. Therefore, I group investments by quarter and model differences in the composition of these groups with covariates. I introduce an aggregated variable "ShareEarlyStage", which equals the share of early stage investments in a quarter, to model different stage compositions. Moreover, I take the mean of all investments' respective industries' performance in the 12 months prior to the investigated quarter as a new covariate. The macroeconomic variables which I use are the same as in the CPV model. Due to the limited number of observations in the fit (76 quarters), I refrain from fitting orthogonal polynomials.[90]

Table 3.5 shows three model specifications of a standard OLS regression.[91] The observations in the regression are weighted with the total amount of capital invested in the respective quarter.[92] The same multi-collinearity issues as in the CPV model require to test the influence of credit spread and IPO market conditions separately from the investment volume of the 12 months before the first investment. Specification 1 shows the results for the credit spread/IPO case. Similar to the CPV case an increasing credit spread and more favorable exit market conditions at the start of the investment lead to a rise in the share of capital that has to be written off. The effect of the growth of industrial production is positive like in the CPV

90 My analysis revealed that the additional degrees of freedom from the orthogonal polynomials are not exploited by the system in the fit, i.e., they do not turn out be significant.

91 I am aware that the shares of defaults in a quarter are truncated figures with a domain of definition between 0 and 1. This would normally speak for applying a Tobit regression model. However, in my case, this effect is negligible. I tested a Tobit model specification but got hardly different results to the OLS specification. To keep the model simple I stick to the OLS regression.

92 All amounts of capital are adjusted for inflation.

Table 3.5: Parameters of the Default Rate of Investments in a Quarter (Credit Risk Evaluation Model)

The table shows the result of three specifications of OLS regressions for a dataset of US VC investments of PE funds between 1985 and 2003. Observations are weighted with the capital which is invested in a quarter. Dependent variable is the percentage share of invested capital in a quarter that is written off completely. Independent variables are macroeconomic parameters of performance indicating investment chances, risks, investment competition and exit market conditions at the time of the first investment into the portfolio company. Variables are as explained in Table 3.1. The indicators *, **, *** signal significance at the 10%, 5% and 1% levels, respectively.

	Dependent variable: % Share of Invested Capital written off in a Quarter					
	Specification 1		**Specification 2**		**Specification 3**	
	Coefficient	z-value	Coefficient	z-value	Coefficient	z-value
Chances						
IndustrialProduction	0.012	1.49	0.004	0.88		
MeanIndustryPerformance	0.001*	1.99	0.001**	2.25	0.001***	3.38
Risks						
CreditSpread	0.169***	5.00				
ShareEarlyStage	0.197	1.25	0.146	1.34		
Competition						
VCInvestmentVolume			0.123***	12.40	0.127***	13.47
Exit market conditions						
IPOs	0.590***	3.85				
ChangeIPOs	0.299	1.29	0.003	0.02		
ChangeM&A	-0.022	-0.78	0.019	0.93	0.026	1.64
(Intercept)	-0.322***	-3.24	-0.122**	-2.00	-0.042	-1.42
Model diagnostics						
Observations	76		76		76	
F-statistic	9.42		63.70		63.70	
p-value	0.00		0.00		0.00	
Adjusted R^2	44.01%		71.34%		71.50%	

model, which is also explicable with the impact of economic cycles on the possibilities to create value in the portfolio companies. The coefficient for IndustryPerformance was negative and not statistically significant in the CPV specifications; in the CREM model MeanIndustryPerformance has positive and statistically significant coefficient. This is in line with expectations: VC investments mostly target innovations within an industry, i.e., new business concepts and new product ideas. These "revolutionary" ideas often change whole industries (e.g., the mobile phone changed the way people communicate and replaced fixed line communication) which potentially leads to a strong stock performance in the industry. If stocks developed very well in the time before the first investment, the probability is high that a major share of an industry change has already happened and is priced in. Thus, a fund that invests into late-moving ventures has a high probability to fail with the investment.

As with the CPV model, the indicator for investment competition is highly statistically significant as expected and explains the majority of the development of defaults. The higher the competition the more investments are written off.

In total, Specification 3 – that is also used for further tests in my prediction model – has a very high adjusted R^2 of 71.5%.

3.4.2.3 Parameters of the Probability Distribution of Non-write-off Investments

The results of my analysis of non-write-off investments are displayed in Table 3.6 for IRRs and in Table 3.7 for MIRRs. As explained in Section 3.3.3.2 I fitted a BCPE distribution to the logarithm of the respective performance measures using GAMLSS. With 2,365 observations in these specifications the multi-collinearity issues between CreditSpread and VCInvestmentVolume that affected estimates in the CPV and CREM models are not significant, but the influence of the level of IPOs may not be included in the analysis.

Well established model diagnostic measures like R^2 or F-statistics are not available in GAMLSS. So, I performed detailed checks of the stability of the model as will be explained in Section 3.4.2.4. Furthermore, diagnosis of the residuals allows to check the fit of the model to the data. Most importantly, the residuals of the fit have to be normally distributed. The fulfilment of this condition is assured in the model diagnostics parts of both tables. Normal distribution of residuals is obtained if the residuals' mean is 0, variance is 1, skewness at 0 and kurtosis at 0 as well. The Filliben correlation coefficient shows the probability that residuals are normally distributed. In my specifications this probability is always 99.9% or even higher.

The GAMLSS displays dependencies for the centrality parameter μ and the dispersion parameter σ of the BCPE distribution. The other two parameters ν and τ were not modeled in dependency of covariates but just fitted with an intercept. As the tables 3.6 and 3.7 show, the Box-Cox-parameter ν is about 2.5 resp. 2.6 for the IRR/MIRR specifications and statistically significant. Thus, a distribution transformation to control for skewness is necessary. A kurtosis adaptation is necessary for the IRR distribution which is indicated by the statistically significant parameter τ. For the MIRR distribution the parameter τ is not statistically significant different from zero.

Concerning the parameters of performance, I fitted orthogonal polynomial splines for some covariates as in the CPV model to control for non-linearities in the relationships (this is indicated with "(poly=Z)" with Z standing for the degree of the orthogonal polynomial). Figures 3.6 and 3.7 show the functional forms of the partial effects of covariates on μ and σ in Specification 2 of Table 3.6.

The partial effect of industry production growth on the BCPE distribution median μ is almost linear, negative and statistically significant. More importantly industry production also affects the probability to realize exceptional returns (indicated by the statistically significant effect on σ): Times with minor growth between 0% and 2% are the best to start an investment. Both negative as well as strong positive growth lead to decreasing probabilities of great returns. Combined with the results from the CPV model the optimum investment time should be when the economy is hardly growing. Anti-cyclical investing seems to pay off.

Concerning the effect of the prior performance of an industry my analysis indicates a generally negative effect on μ, which is in line with expectations. Interestingly, the effect is even slightly positive around zero.[93] Thus,

93 As the covariate is used in a logarithmic transformed way, zero growth lies at 4.6, which equals log(100).

Table 3.6: Parameters of Investment Performance of Non-Write Off Investments Measured in IRR

The table shows the result of two specifications of GAMLSS regression models for a dataset of US VC investments of PE funds between 1985 and 2003. Dependent variable is the logarithm of 100 plus an investment's IRR (in %); observations are weighted with the relative share of capital which is invested. Independent variables are macroeconomic parameters of performance indicating investment chances, risks, investment competition and exit market conditions at the time of the first investment into the portfolio company. To control for dependencies from specifics of the investing fund manager and fund, control variables FundSize and FundNumber are included in the specifications. Variables are as explained in Table 3.1. Some covariates are fitted with orthogonal polynomials which is indicated by "(poly=Z)"; Z stands for the orthogonal polynomial of the Zth degree. The indicators *, **, *** signal significance at the 10%, 5% and 1% levels, respectively.

	Dependent variable: log (100+IRR (in %))							
	Specification 1				Specification 2			
	μ		σ		μ		σ	
Link function	Identity link		Log link		Identity link		Log link	
	Coefficients	t-value	Coefficients	t-value	Coefficients	t-value	Coefficients	t-value
Chances								
IndustrialProduction (poly=1)	-0.83*	-1.71	-5.22***	-3.00	-0.01***	-2.89	-6.38***	-4.10
IndustrialProduction (poly=2)	-0.67	-1.40	-6.79***	-5.60			-7.61***	-6.37
IndustryPerformance (poly=1)	0.90	1.64	0.63	0.45	0.68	1.27		
IndustryPerformance (poly=2)	-1.88***	-3.53	-0.77	-0.63	-1.43***	-2.65		
IndustryPerformance (poly=3)	-3.71***	-9.50			-4.42***	-10.05		
Risks								
CreditSpread (poly=1)	5.58***	9.29	-10.52***	-5.20	4.14***	7.05	-11.06***	-5.69
CreditSpread (poly=2)	4.02***	9.14	-2.89**	-2.15	4.24***	9.79	-3.88***	-3.26
LaterStage	0.14***	7.49	-0.18***	-3.71	0.14***	7.83	-0.21***	-4.30
UnspecifiedVC	0.00	0.04	-0.02	-0.28	-0.01	-0.17	-0.04	-0.53
Industrials	0.02	0.81	-0.22**	-2.02	0.03	1.34	-0.24**	-2.21
Computer	-0.15***	-4.55	0.35***	4.48	-0.13***	-4.05	0.31***	4.01
Software	-0.05*	-1.65	0.19**	2.44	-0.07**	-2.17	0.18**	2.34
Internet&Media	-0.22***	-7.81	0.27***	2.88	-0.17***	-5.87	0.24***	3.02
Telecom	0.05	1.57	0.29***	3.72	-0.04	-1.46	0.30***	3.97
Other	-0.07***	-3.35	-0.04	-0.41	0.00	-0.15	-0.10	-1.10
Competition								
VCInvestmentVolume (poly=1)	-13.24***	-23.35	28.40***	15.34	-12.76***	-22.94	29.77***	16.78
VCInvestmentVolume (poly=2)	-1.46***	-2.73	-12.83***	-9.89	-0.90*	-1.76	-14.52***	-11.82
VCInvestmentVolume (poly=3)	1.90***	4.52	-6.94***	-4.60	2.27***	5.41	-7.25***	-5.27
Exit market conditions								
ChangeIPOs (poly=1)	-0.50***	-2.81	-4.06***	-2.80	-0.62***	-3.52	-1.64***	-3.25
ChangeIPOs (poly=2)			-0.08	-0.06				
ChangeM&A (poly=1)	0.19	0.32	-0.08	-1.55				
ChangeM&A (poly=2)	0.45	0.64						
Control variables								
DivestBubble			0.43***	5.39			0.40***	5.02
FundSize (poly=1)	0.21	0.29	-9.85***	-5.18	0.40	0.56	-0.18***	-8.49
FundSize (poly=2)	-1.44***	-2.89	-2.13	-1.47	-1.65***	-3.15		
FundNumber (poly=1)	-0.56	-0.93	1.16	0.93				
FundNumber (poly=2)	0.02	0.04	-0.05	-0.05				
(Intercept)	4.61***	210.35	-1.78***	-28.56	4.64***	194.52	1.74***	4.10
Box-Cox-Parameters								
ν	2.48***	30.06			2.60***	32.30		
τ	-0.08**	-2.48			-0.09***	-2.61		
Observations	2,365				2,365			
Model diagnostics (quantile residuals)								
mean	-0.02				-0.02			
variance	1.07				1.06			
coef. of skewness	-0.04				-0.06			
coef. of kurtosis	0.04				0.04			
Filliben correlation coefficient	99.9%				100.0%			

Figure 3.6: Functional Form of Partial Effects of Covariates on μ in Specification 2 of Table 3.6

The vertical axis of the charts plots the partial effects of the covariates. Dotted lines indicate standard errors.

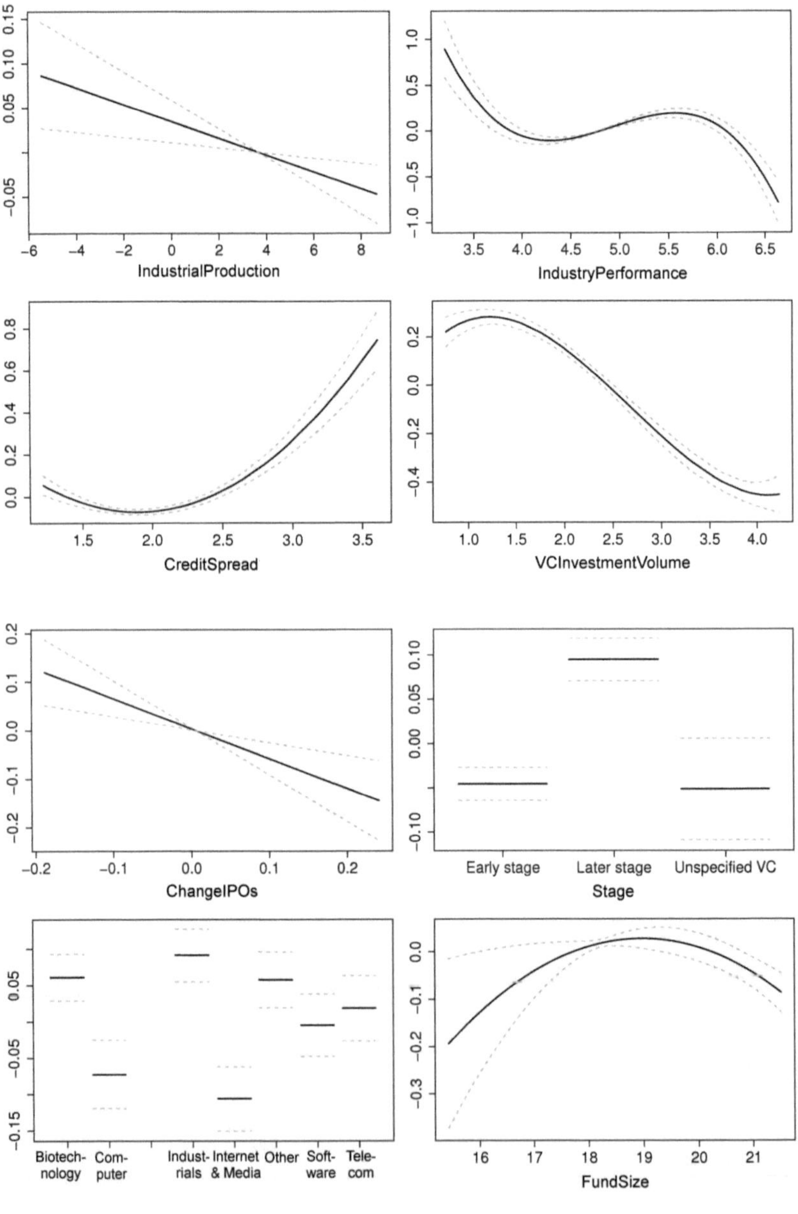

Figure 3.7: Functional Form of Partial Effects of Covariates on σ in Specification 2 of Table 3.6

The vertical axis of the charts plots the partial effects of the covariates. Dotted lines indicate standard errors.

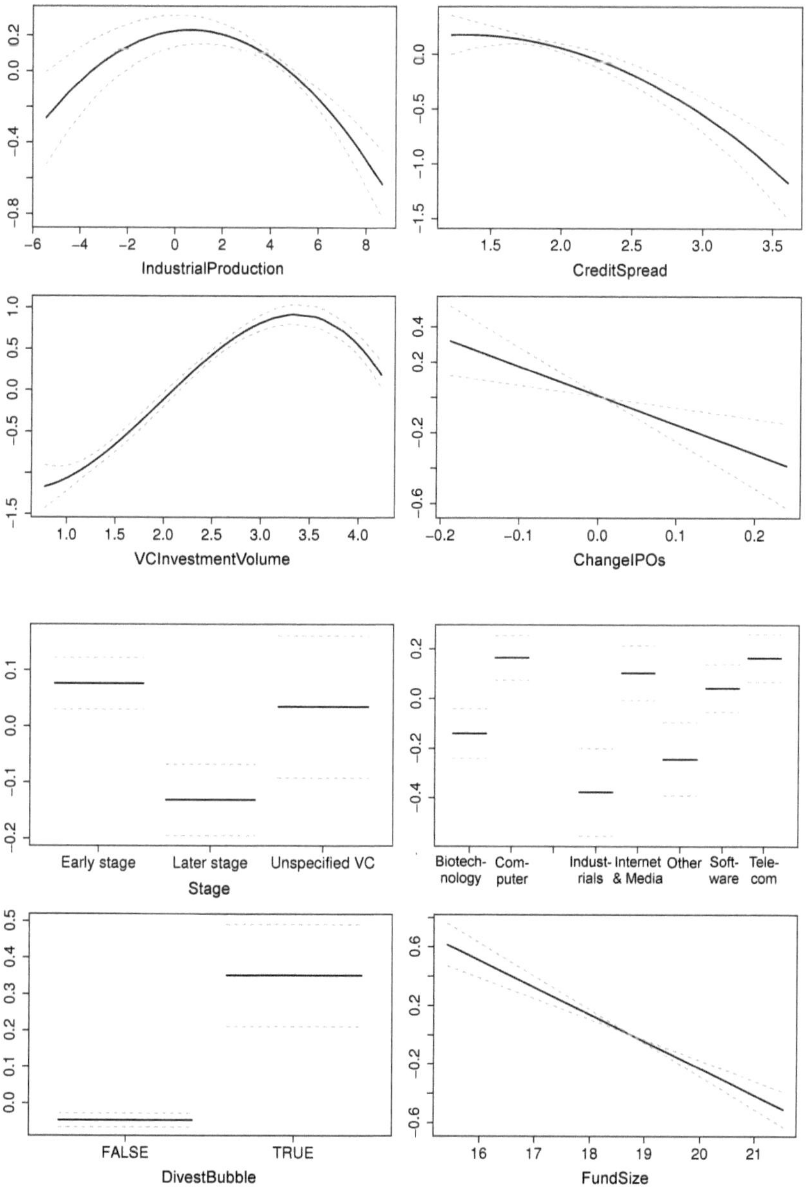

Table 3.7: Parameters of Investment Performance of Non-Write-Off Investments Measured in MIRR

The table shows the result of two specifications of GAMLSS regression models for a dataset of US VC investments of PE funds between 1985 and 2003. Dependent variable is the logarithm of 100 plus an investment's MIRR (in %); observations are weighted with the relative share of capital which is invested. Independent variables are macroeconomic parameters of performance indicating investment chances, risks, investment competition and exit market conditions at the time of the first investment into the portfolio company. To control for dependencies from specifics of the investing fund manager and fund, control variables FundSize and FundNumber are included in the specifications. Variables are as explained in Table 3.1. Some covariates are fitted with orthogonal polynomials which is indicated by "(poly=Z)"; Z stands for the orthogonal polynomial of the Zth degree. The indicators *, **, *** signal significance at the 10%, 5% and 1% levels, respectively.

	Dependent variable: log (100+MIRR (in %))							
	Specification 1				Specification 2			
	μ		σ		μ		σ	
Link function	Identity link		Log link		Identity link		Log link	
	Coefficients	t-value	Coefficients	t-value	Coefficients	t-value	Coefficients	t-value
Chances								
IndustrialProduction (poly=1)	0.09	0.23	-1.52	-0.92			-0.98	-0.61
IndustrialProduction (poly=2)	-0.14	-0.35	-4.31***	-3.75			-4.31***	-3.99
IndustryPerformance (poly=1)	-0.20	-0.43	1.24	0.95	0.16	0.35		
IndustryPerformance (poly=2)	-0.74*	-1.65	-0.79	-0.68	-0.89**	-2.00		
IndustryPerformance (poly=3)	-2.53***	-6.92			-2.65***	-6.79		
Risks								
CreditSpread (poly=1)	2.86***	5.64	-4.26**	-2.22	3.23***	7.06	-3.93**	-2.27
CreditSpread (poly=2)	1.87***	5.66	-3.89***	-3.05	2.25***	7.03	-4.42***	-3.88
LaterStage	0.09***	6.70	-0.11**	-2.35	0.09***	6.53	-0.09*	-1.92
UnspecifiedVC	0.02	0.61	0.08	1.01	-0.02	-0.69	0.10	1.36
Industrials			-0.14	-1.43			-0.17*	-1.67
Computer			0.16**	2.17			0.14**	1.98
Software			0.13*	1.72			0.11	1.44
Internet&Media			0.25***	2.77			0.24***	3.09
Telecom			0.32***	4.42			0.29***	4.03
Other			-0.11	-1.21			-0.13	-1.44
Competition								
VCInvestmentVolume (poly=1)	-8.29***	-17.31	21.94***	12.60	-8.75***	-19.95	22.11***	13.84
VCInvestmentVolume (poly=2)	-0.77*	-1.91	-11.02***	-8.97	-0.16	-0.46	-11.86***	-10.10
VCInvestmentVolume (poly=3)	0.91**	2.54	-5.73***	-4.00	1.21***	3.93	-4.84***	-4.40
Exit market conditions								
ChangeIPOs (poly=1)	-0.48***	-3.15	-1.00	-0.73				
ChangeIPOs (poly=2)			0.08	0.07				
ChangeM&A (poly=1)	0.02	0.03	-0.10**	-2.16			-0.12***	-2.76
ChangeM&A (poly=2)	-0.21	-0.40						
Control variables								
DivestBubble			0.52***	6.90			0.53***	7.07
FundSize (poly=1)	-0.19	-0.31	-0.15***	-6.67	0.21	0.39	-0.16***	-8.04
FundSize (poly=2)	-1.31***	-3.19			-1.59***	-3.89		
FundNumber (poly=1)	0.73	1.57	-0.46	-0.39				
FundNumber (poly=2)	-0.52	-1.29	1.32	1.27				
(Intercept)	4.63***	365.90	0.54	1.25	4.63***	371.28	0.82**	2.06
Box-Cox-Parameters								
ν	2.61***	19.46			2.64***	19.95		
τ	0.02	0.63			0.03	0.80		
Observations	2,365				2,365			
Model diagnostics (quantile residuals)								
mean	-0.01				-0.02			
variance	1.13				1.13			
coef. of skewness	0.01				0.02			
coef. of kurtosis	0.30				0.30			
Filliben correlation coefficient	99.9%				99.9%			

a slightly positive growth is better than a slightly negative growth before the first financing round of a portfolio company. Though, heavy prior growth of a sector reduces value generation opportunities dramatically.

The partial effects of the credit spread in the GAMLSS specifications confirm my hypothesis that a rise in market risks increases the fund managers' risk aversion; they invest more in stable companies. On the one hand, this decreases the probability of exceptional performing investments but increases the median of the log-IRR distribution statistically significant.

Unsurprisingly, early stage investments are more volatile than later stage investments and have a lower median log-IRR. Investments into different sectors hardly differ concerning their median, but do so concerning the dispersion of returns. Investments into the sectors "Industrials" and "Other" have a lower performance distribution spread, all other sectors a higher spread than the reference sector "Biotechnology".

Competition between funds at the time of investment heavily affects the shape of the performance distribution pattern. The impact is almost linear negative for μ. As one can see in the CPV and CREM specifications, hot market phases with strong competition lead to massive write-offs. At the same time, the chance for exceptional returns increases. On the first sight, it might surprise that this increase is not linear and tips beyond a certain point. This result may be explained with the fact, that extremely competitive environments, like in the year 1999, historically occurred shortly before heavy market downturns which destroyed chances to divest successfully.

Positive changes in the condition of the IPO market negatively affect both μ and σ of the return distribution in line with what I hypothesized. This may be explained by the cycle theory: When fund managers invest in a time the IPO market picks up the probability is high that conditions are bad at the moment they want to divest. On the contrary, investments bought during worsening exit market conditions may profit from a later economic recovery and may be sold at favorable exit market conditions.

Although experience (measured with FundNumber) hardly seems to matter, the size of the VC fund affects investment success. In the CPV model, I found out that a fund size of about USD 300M is optimal for investing. This result is confirmed by the analysis of the effect of fund size on μ. Concerning the partial effect of size on σ I find that funds bigger than USD 300M include less super-performing investments, while smaller funds generate even slightly more. These effects are in line with results from Kaplan and Schoar (2005, p. 1804) who find a concave relationship for fund size; for the fund number they find a positive linear influence.

Summarizing,the results in Table 3.7 for MIRR are very similar to the results for IRR. This confirms the stability of my results.

3.4.2.4 Robustness Checks

To check the robustness of my results I performed detailed analysis on several levels:

- I analyzed the performance of VC investments with the two different performance measures IRR and MIRR which revealed similar parameters of performance.
- The covariates - especially the macroeconomic variables - are intercorrelated. This may lead to multi-collinearity problems unbalancing estimates and leading to wrong conclusions concerning significance of covariates. In the CPV and the CREM, I checked the Variance Inflation Factor (VIF) of the covariates and both models did not indicate multi-collinearity problems. For the GAMLSS model methods to check multi-collinearity problems are not yet developed. This is why I Box-Cox-transformed the dependent variables according to the results of the GAMLSS and fitted a linear model similar to the GAMLSS specification for the centrality parameter μ and controlled VIFs. Although the linear model does not fully control for heterogeneity as the GAMLSS, I see this approach as a possible way to get an indication on multi-collinearity problems. In my specifications, VIFs of the linear model indicated no multi-collinearity issues for the non-write-off models.

- The GAMLSS assumes that the probability distribution of the dependent variable follows a given pattern. To make sure that the distribution pattern I use is the best approximation to the data, I tested a long list of distributions including the Log-Normal distribution used by other authors. My analysis revealed that for a viable quality of the prediction model, a distribution pattern like the BCPE distribution is required which allows modeling skewness and kurtosis.
  The functional form of the partial effects of the investigated covariates on location and scale parameters μ and σ in the GAMLSS were quite similar across the specifications of different distribution patterns[94]. To compare different specifications and check the quality of the fit of different distribution patterns I investigated the models' AICs and SBCs as well as the shape of their residuals. The closer the distribution of the residuals is to a normal distribution, the better is the approximation of the model. As Tables 3.6 and 3.7 show, fitting a BCPE distribution to the data results in almost perfectly normal distributed residuals in my specifications.
- For many of the covariates in my models I use different functional forms than linearity (i.e., orthogonal polynomials with quadratic or cubic shapes). To come up with these specifications I used the following methodology: First, I analyzed scatter plots of the interrelation between the covariates and the performance measures as well as between the covariates and the dispersion of performance measures. I compared these results with the results of General Additive Models fitted with splines to the data and decided upon a first specification to test in the GAMLSS. After incorporating the covariates with non-linear functional forms in the GAMLSS I tested for the need of more than the first degree of the polynomial and excluded second and/or third degree polynomials depending on their statistical significance. Afterwards, I substantiated the results fitting different time frames and leaving out observations of different years (see Section 3.4.3).

94 Sometimes covariates were not significant.

Moreover, I fitted the same specifications without weighting the observations with the investment size. Although not all covariates were statistically significant in all specifications, the functional forms were confirmed throughout the fits.

3.4.3 Power of the Prediction of the Direct Investment Return Distribution Pattern

The power of my model may be ultimately tested by assessing its fit over time and its' predictive capabilities. I perform several tests. First, I investigate how well the model approximates the data overall. As Figure 3.8 shows, the mean default rate in the data sample equals almost exactly the predicted rate. Also the shape of the fitted BCPE distribution approximates the shape of the IRR distribution pattern very well.

Figure 3.8: Overall Fit of the Predicted Distribution Pattern to the Real Distribution Pattern of IRRs for All Observations in Sample

The displayed probability distribution in this figure is the weighted mean of the predicted probability distributions of individual investments. This means for example that the probability that an IRR is below 0% equals the weighted mean of the predicted probabilities of all investments that their respective IRR is below 0%. Single investments' predictions are weighted with the relative share of capital which is invested.

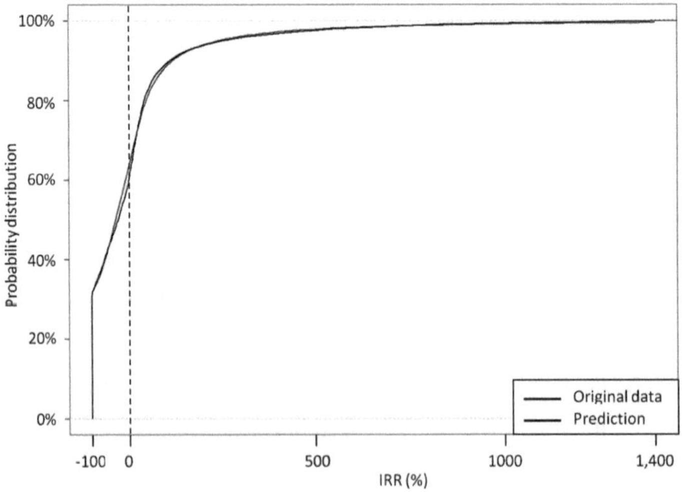

Table 3.8: Comparison of results of my prediction approaches (CPV based and CREM based) with results based on Buchner at al. (2009)

The table shows a comparison of the precision of different prediction methodologies along the four measures "share of defaults", "share of super returns", "median IRR" and "mean IRR". I use IRR distributions in different investment years for the calculations. The rows marked with "original data" shows the descriptive statistics on my CEPRES sample. The rows marked with "All data considered" shows the results in case all sample observations are used to develop the prediction models. The rows marked with "Data of predicted year removed" shows the results in case all observations except from the observations of the year, which I want to predict, are used to develop the prediction models.

Data used for prediction	Methodology	1986	1987	1988	1989	1990	1991	1992	1993	1994	1995	1996	1997	1998	1999	2000	2001	2002	2003	Avg. error predicted vs. original	Std. error predicted vs. original
Share of defaults (%)																					
	Original data	20	13	18	18	12	7	9	7	18	25	19	26	31	42	48	41	29	41		
All data considered	Buchner et al.	32	32	32	32	32	32	32	32	32	32	32	32	33	35	35	34	34	34	**-9**	**15**
	CPV based	20	20	17	15	13	10	16	19	19	20	25	29	34	43	49	37	36	33	**-2**	**5**
	CREM based	13	17	16	15	12	7	9	15	14	20	24	30	33	41	47	41	32	29	**0**	**5**
Data of predicted year removed	Buchner et al.	32	33	33	33	33	32	33	33	33	33	33	33	34	32	30	34	34	33	**-9**	**16**
	CPV based	20	21	16	15	13	10	16	20	19	20	26	29	34	40	52	34	42	31	**-2**	**7**
	CREM based	13	18	14	18	8	7	9	16	13	19	25	31	35	37	46	45	33	28	**0**	**9**
Share of super returns (%)																					
	Original data	19	1	4	4	9	10	11	5	10	15	17	22	20	8	1	1	7	24		
All data considered	Buchner et al.	12	12	12	12	12	12	12	13	12	12	12	12	11	11	11	11	11	11	**-1**	**7**
	CPV based	11	5	6	15	10	9	7	7	9	14	15	13	15	11	4	6	11	13	**0**	**5**
	CREM based	12	5	6	15	10	9	7	8	9	14	15	13	15	11	4	6	12	13	**0**	**5**
Data of predicted year removed	Buchner et al.	12	12	12	12	12	13	12	13	12	12	12	11	10	11	13	12	11	11	**-1**	**8**
	CPV based	10	5	6	16	12	6	6	8	8	14	16	13	13	14	5	10	12	12	**0**	**6**
	CREM based	11	6	6	15	12	6	7	8	8	14	17	12	13	14	6	8	13	12	**0**	**5**
Median IRR																					
	Original data	18.4	22.7	9.7	10.0	19.0	31.7	18.4	15.2	10.1	3.8	21.3	2.0	-26.0	-78.5	-92.6	-63.8	-33.3	-49.0		
All data considered	Buchner et al.	-10.3	-14.7	-10.7	-13.1	-8.3	-2.6	-5.1	-1.9	-8.5	-6.3	-8.1	-14.8	-35.7	-48.7	-53.0	-45.4	-38.6	-39.5	**11.3**	**24.2**
	CPV based	17.7	12.9	7.3	19.2	21.8	29.9	17.6	12.1	12.6	9.7	-5.1	-19.7	-40.2	-77.1	-92.3	-49.8	-39.0	-30.7	**1.8**	**11.0**
	CREM based	23.1	14.9	8.1	19.2	22.2	31.3	20.8	14.8	15.9	10.5	-4.0	-20.6	-38.3	-71.4	-87.7	-57.8	-31.2	-22.3	**-0.5**	**11.5**
Data of predicted year removed	Buchner et al.	-10.9	-15.3	-11.6	-15.6	-9.5	-4.4	-5.2	-1.7	-9.6	-8.1	-11.9	-14.4	-45.7	-42.2	-38.6	-45.1	-45.0	-38.4	**11.8**	**27.4**
	CPV based	18.7	6.1	7.7	19.7	22.1	23.9	17.2	10.8	9.7	13.2	-11.3	-17.7	-41.1	-60.9	-100.0	-54.3	-60.6	-26.3	**3.4**	**15.0**
	CREM based	23.6	8.5	8.7	17.3	25.0	25.6	20.4	13.7	13.5	13.6	-9.5	-20.8	-43.6	-53.0	-77.6	-78.5	-38.0	-21.8	**0.7**	**17.8**
Mean IRR																					
	Original data	21.1	11.7	-2.2	5.3	11.8	35.3	21.5	25.1	24.0	10.3	30.3	39.6	66.5	-23.6	-61.6	-44.6	-12.7	-3.7		
All data considered	Buchner et al.	13.3	10.2	12.2	10.4	14.6	16.9	15.5	16.6	14.0	15.2	15.1	10.0	-3.4	-9.2	-11.4	-7.5	-3.2	-5.2	**1.7**	**24.8**
	CPV based	16.9	4.8	3.7	30.6	23.2	29.1	11.7	8.3	10.9	20.9	20.8	8.4	8.6	-17.4	-51.7	-31.8	-9.2	0.4	**3.7**	**18.8**
	CREM based	27.4	8.6	5.2	30.6	24.0	33.2	20.3	13.7	17.3	21.9	22.0	7.6	10.0	-14.1	-50.5	-35.8	-3.8	6.9	**0.5**	**18.3**
Data of predicted year removed	Buchner et al.	13.0	10.0	12.5	9.2	13.8	16.0	15.5	16.2	13.7	12.6	11.9	7.6	-13.1	-6.6	3.3	-5.5	-6.3	-4.9	**1.9**	**28.6**
	CPV based	16.2	-0.5	4.6	34.0	26.1	21.2	10.3	7.9	7.3	23.5	23.1	5.5	0.1	1.2	-47.4	-17.4	-11.0	-1.1	**2.8**	**23.0**
	CREM based	26.3	4.2	6.9	29.8	32.1	25.9	19.4	13.2	15.2	24.1	24.9	2.8	-1.6	6.8	-40.9	-31.5	2.4	2.8	**-0.5**	**18.8**

Second, I test how well my model may predict the distributions of IRRs in single years between 1986 and 2003.[95] Therefore, I compare my predictions with the predictions of the model closest related to mine from Buchner et al. (2010) who have the same intensions as I have. The goal is to prove the superiority of my approach. I replicated the authors' model including the OLS regression specification which they use for their predictions.[96] I apply their approach to my data sample and – unsurprisingly – find very similar results to the ones illustrated in Buchner et al. (2010) applying their regression specification.[97]

To judge how close the predicted distributions of IRRs over time are to the real distributions I use four measures: the share of defaults (IRR=-100%), the share of super returns (IRR>=100%), the median and the mean of the distributions. I compare the original IRR distributions from the sample, the predicted distributions based on my approaches (CPV based and CREM based) and the predicted distributions by the methodology of Buchner et al. (2010) along the four measures (see Table 3.8). Therefore, I perform two calculations: In the first calculation, I use all observations of my sample to determine the parameters in the respective regression models and then predict the IRR distributions for the years 1986 to 2003 based on these models. Because the observations for which I want to make predictions are included in the model determination, I automatically overestimate the quality of the models. This is why I exclude the observations, which I want to predict, in the regression analysis for the second set of calculations.

As Table 3.8 shows, my methodology is superior in every respect to the methodology suggested by Buchner et al. (2010): In the case that all observations are used for determining the prediction model, the difference

95 The year 1985 was not considered as I do not have a sufficiently high number of data points to assess the probability distribution of returns in 1985.

96 My model is different to the specification of Buchner et al. (2010) just in one respect: I didn't include a dummy variable which indicates that a portfolio company is from the US as all my portfolio companies are from the US.

97 I do not show the results here. They may be received upon request.

between the real and the predicted default rate is -9% in the investigated 18 years on average (i.e., the mean of all differences) with a 15% standard deviation (i.e., the square root of the squared differences). This result based on the model of Buchner et al. (2010) compares to an average error of -2% (0%) and a standard error of 5% (5%) with my CPV (CREM) based model.

My models predict the share of super returns better as well. The average deviation in the predicted shares is 0% with a standard deviation of 5% for my models. The methodology of Buchner et al. has an average error of -1% and a higher standard deviation at 7%. The superior precision of my model implies a good approximation of my models' median returns to the real median over time. The average error in the median prediction is just 1.8% (-0.5%) for the CPV (CREM) based approaches and the standard error is 11.0% (11.5%). This compares to an average error of 11.3% and a standard error of 24.2% in the median prediction by Buchner et al. (2010).

With the high skewness of the IRR distribution the mean is quite unreliable for the comparison of a predicted distribution with the real distribution. However, for reasons of completeness I also consider the mean of the IRR distribution. The mean IRR prediction by the methodology from Buchner et al. (2010) has a lower average error than the error with my CPV based approach; but both predictions with my two approaches (CPV and CREM based) are less dispersed than the predictions from Buchner et al. (2010).

The results of the predictions, which do not include the observations to be predicted in the regression analysis to determine the parameters, confirm the superiority of my approaches across all four measures which I applied.

In conclusion, all four measures "share of defaults", "share of super returns", "median IRR" and "mean IRR" point to a more precise prediction with my methodologies than with the methodology from Buchner et al. (2010) over time.

Concerning the comparison of the CPV based approach with the CREM based approach my results in Table 3.8 are not unambiguous.

Overall, the combined models consisting of a GAMLSS model and a CPV resp. CREM model are very supportive to estimate the effects of the macroeconomic environment on the probability distribution patterns of VC direct investments.

3.5 Extensions of the Model to Predict the Performance Distribution Pattern of Funds and Funds-of-Funds

The prediction of direct VC investment returns is interesting both for VC fund managers and for VC investors to time their investments. Though, VC investors are also interested in predicting the returns of whole VC funds they think about investing in. My methodology described in previous chapters may be extended to the prediction of **fund** returns, i.e., the performance of portfolios of direct investments. Generally two approaches seem applicable: The first approach (Section 3.5.1) is an adaptation of an existing methodology, the second approach (Section 3.5.2) is new and promises superior results.

3.5.1 Approach Using the Mean IRR of a Fund's Direct Investments

The aggregation of investment returns of single investments of a fund to the performance distribution pattern of the whole fund is not as simple as it seems. In Section 3.3.2, I mentioned that the IRR of a (pooled) portfolio of investments does not equal the arithmetic average of the individual investments' IRRs. Especially high IRR values of single investments distort fund IRRs that are calculated as the average of the investment IRRs. Nevertheless, the arithmetic average seems a good starting point and is also used by Buchner et al. (2010) for their fund return prediction: In a first step, Buchner et al. use a Monte Carlo Simulation (MCS) to predict the returns of a predefined number of single investments. In a second step, they calculate the mean IRR in the portfolio. When the two steps are reiterated a high number of times drawing lots from the predicted return distributions of the direct investments, one gets an approximation of the predicted portfolio IRR distribution pattern.

The integration of my direct investment return prediction approach into the fund return prediction methodology of Buchner et al. (2010) is straight forward: One needs to find reasonable assumptions for the development of the macroeconomic environment in the fund's investment period (usually 4-6 years) and for the shares of the committed capital that are invested in respective fund years (for a possible solution to this question see Buchner et al., 2008). These shares are required to weight investment IRRs in the average calculation. As predictions of macroeconomic developments are arguable, one may develop scenarios and weight them. Based on the assumptions of future economic development, my approach – as described in Section 3.3.3 – allows to predict the investment year IRR distribution patterns over the fund's investment period. For the combination of these distribution patterns to a fund IRR, one has to draw lots from the yearly distributions and calculate the weighted mean. Frequent repeats of this calculation bring about the desired continuous gross fund return distribution pattern. To get the net returns, fund management fees have to be deducted from the gross returns in dependency from the fund's fee structure.

An analogous procedure may be used to calculate fund-of-fund returns.

3.5.2 Approach Using the Mean MIRR of Direct Investment Returns over a Fund's Lifetime

A better methodology than the one just described relies on the MIRR return measurement concept: From the perspective of an investor into a VC fund – who is the key interest group of a fund return distribution – it is especially relevant how a fund will perform in the future. Due to the "false" reinvestment hypothesis in the IRR calculation (see Section 3.3.2), the IRR figure is less meaningful than the MIRR which avoids this hypothesis. Thus, predicting the fund return distribution pattern measured in MIRR should be in the investor's interest.

The calculation of the fund MIRR distribution pattern is more complicated than calculating a mean IRR, but numerically stable and not skewed by high individual investments. To determine the MIRR return distribution pattern one requires:

1. the timing of investments within the investment period of the fund,
2. the respective invested amounts or shares of the respective amounts relative to the fund size,
3. the annual returns of the respective investments and
4. the investment duration of the respective investments.

The return distribution and the investment duration depend on the macroeconomic situation at the time of investment. Thus, the future macroeconomic development needs to be estimated for the investment period of the fund.[98] Moreover, a discount rate (see Section 3.3.2) is required for the MIRR calculation which may be chosen arbitrarily.

Figure 3.9 illustrates the suggested new approach to predict the fund MIRR distribution; it has the following steps:

1. Predict the development of the macroeconomic environment x_t for the years $t = 1, ..., I$ in the fund investment period I.
2. Predict the MIRR distributions g_t of the investments in the different investment years t using the methodology described in Section 3.3.3 under consideration of the predicted macroeconomic parameters x_t.
3. Predict the investment duration distribution d_t for the different investment years based on the expected MIRR distribution g_t and macroeconomic factors x_t (how to do this will be described after this enumeration).
4. Calculate the multiple u_t which investments in individual investment years generate over the fund's whole lifetime T. Therefore, one has to raise $(1+g_t)$ by the power of d_t for every t respectively and compound the result for the rest of the fund's lifetime $T - d_t$ with the alterna-

98 Based on the methodology that I suggested in Section 3.3.3 it is just necessary to project the future macroeconomic development for the investment period and not for the whole maturity of the fund. This is an advantage in comparison to the methodology suggested by Buchner et al. (2010).

Figure 3.9: Illustration of Approach to Predict the Return Distribution of a VC Fund based on Individual Investments

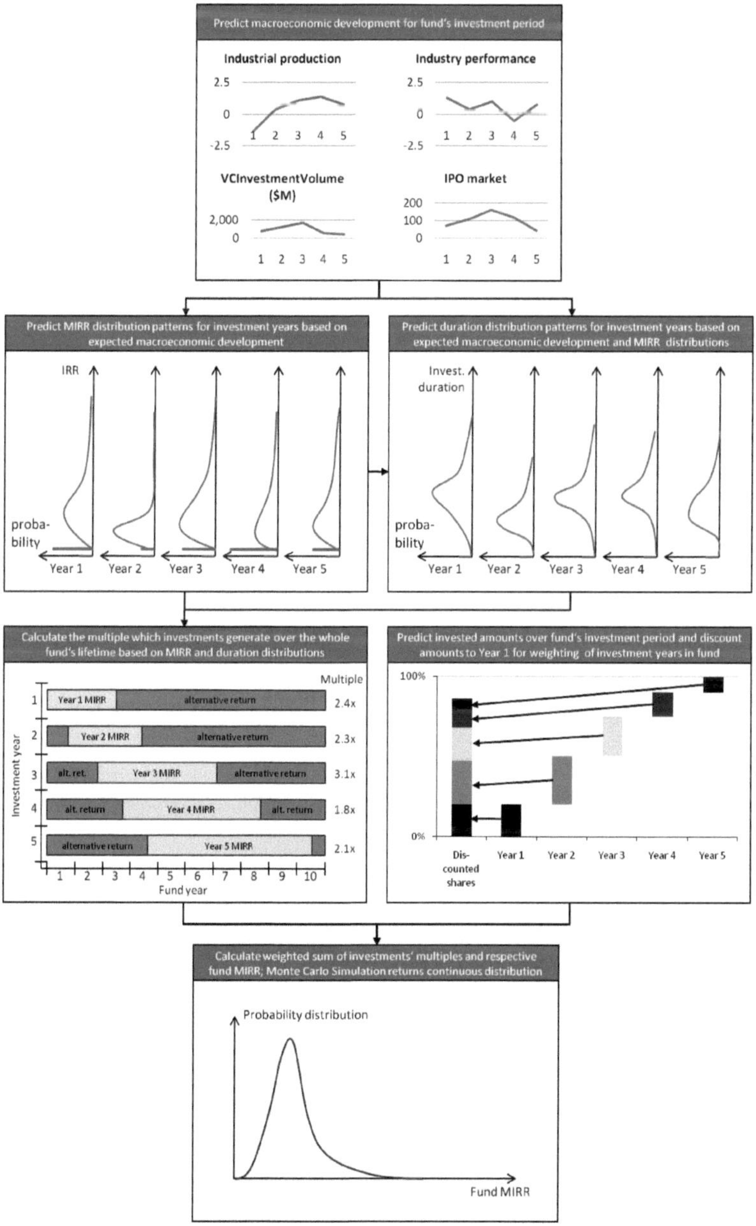

tive rate $1 + r$ (which is the same rate as the one uses in the MIRR calculation), i.e.,

$$u_t = (1 + g_t)^{d_t} \cdot (1 + r)^{T-d_t}. \tag{3.17}$$

5. Make assumptions on the share of committed capital h_t invested in the respective years of the investment period ($\sum_{t=1}^{I} h_t = 1$) and discount them with the alternative rate $1 + r$ to the fundraising year of the fund, i.e., $\frac{h_t}{(1+r)^{t-1}}$. A possible methodology to assess the annual shares is suggested by Buchner et al. (2008).[99]
6. Calculate the weighted sum of the investment year multiples u_t (weighted by $\frac{h_t}{(1+r)^{t-1}}$) which equals the fund multiple U, i.e.,

$$\begin{aligned} U &= \sum_{t=1}^{I} \frac{h_t}{(1+r)^{t-1}} \cdot u_t && (3.18) \\ &= \sum_{t=1}^{I} \frac{h_t}{(1+r)^{t-1}} \cdot \left[(1 + g_t)^{d_t} \cdot (1 + r)^{T-d_t}\right]. && (3.19) \end{aligned}$$

7. Calculate the annual MIRR function G by extracting the Tth root of the fund multiple U. Thus, the fund MIRR may be written as:

$$G = \sqrt[T]{U} = \sqrt[T]{\sum_{t=1}^{I} \frac{h_t}{(1+r)^{t-1}} \cdot [(1 + g_t)^{d_t} \cdot (1 + r)^{T-d_t}]}. \tag{3.20}$$

8. Reiterate steps 1-7 a high number of times (e.g., 10,000 times) for different lots from the return distributions g_t and from the corresponding duration distributions d_t to get a continuous fund MIRR distribution G. This may be done with a MCS.

A precondition for the suggested approach is that the duration distribution pattern d_t may be well predicted. In order to do so, I investigated what

99 The approach of Buchner et al. (2008) does not consider the effect of macroeconomic factors on the investment policy of a fund. Although this could be considered in addition, I do not focus on this aspect in the current work and thus leave it open for future research.

Table 3.9: Parameters of Investment Duration of Non-Write-Off Investments
The table shows the result of two specifications of GAMLSS regression models for a dataset of US VC investments of PE funds between 1985 and 2003. Dependent variable is the logarithm of the investment duration (in months); observations are weighted with the relative share of capital which is invested. Independent variables are the logarithm of an investment's MIRR and macroeconomic as well as investment specific parameters indicating investment chances, risks and investment competition at the time of the first investment into the portfolio company. Variables are as explained in Table 3.1. Some covariates are fitted with orthogonal polynomials which is indicated by "(poly=Z)"; Z stands for the orthogonal polynomial of the Zth degree. The indicators *, **, *** signal significance at the 10%, 5% and 1% levels, respectively.

	Dependent variable: log(Investment Duration (in months))			
	μ		σ	
Link function	Identity link		Log link	
	Coefficients	t-value	Coefficients	t-value
log(MIRR) (poly=1)	-5.60***	-15.36		
log(MIRR) (poly=2)	-8.08***	-25.84		
Chances				
IndustrialProduction (poly=1)	2.04***	5.05	-1.34**	-1.97
IndustrialProduction (poly=2)	1.16***	3.14	-3.34***	-5.21
Risks				
LaterStage	-0.04**	-2.22		
UnspecifiedVC	-0.13***	-5.85		
Competition				
VCInvestmentVolume (poly=1)	-9.38***	-22.70	1.82***	2.88
VCInvestmentVolume (poly=2)	3.19***	7.33	-2.28***	-3.31
Control variables				
DivestBubble	-0.11***	-4.32		
(Intercept)	4.01***	342.77	-2.13***	-166.97
Box-Cox-Parameters				
ν	3.42***	29.41		
τ	0.95***	19.37		
Observations	2,365			

Model diagnostics (quantile residuals)	
mean	0.00
variance	1.01
coef. of skewness	-0.04
coef. of kurtosis	0.13
Filliben correlation coefficient	100.0%

Figure 3.10: Functional Form of Partial Effects of Covariates on μ in Table 3.9
The vertical axis of the charts plots the partial effects of the covariates. Dotted lines indicate standard errors.

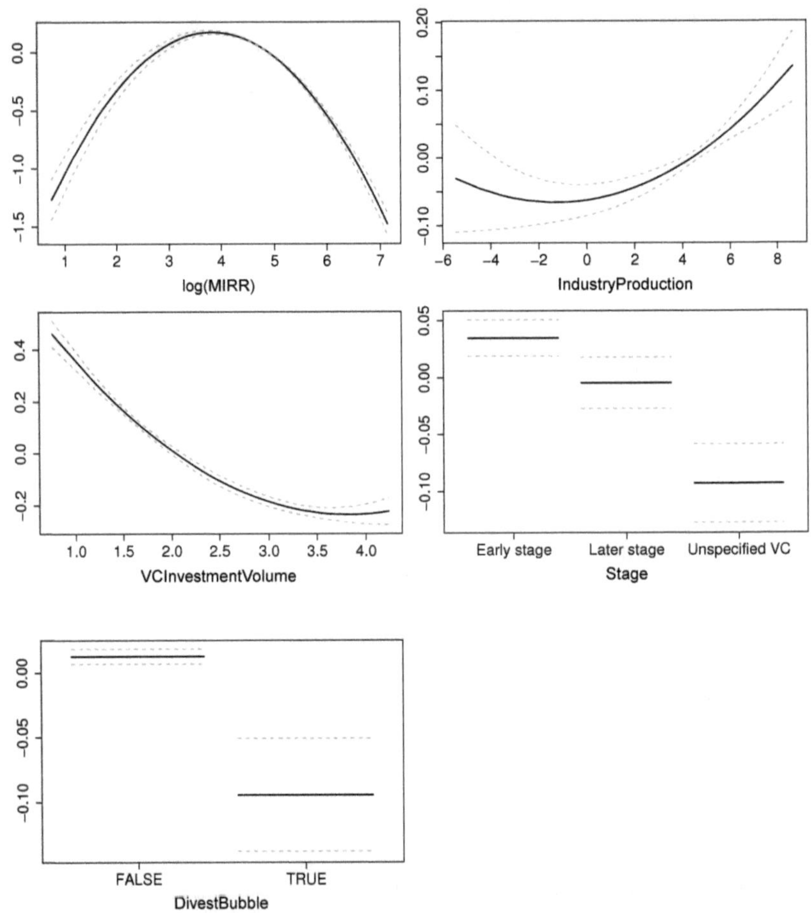

Figure 3.11: Functional Form of Partial Effects of Covariates on σ in Table 3.9
The vertical axis of the charts plots the partial effects of the covariates. Dotted lines indicate standard errors.

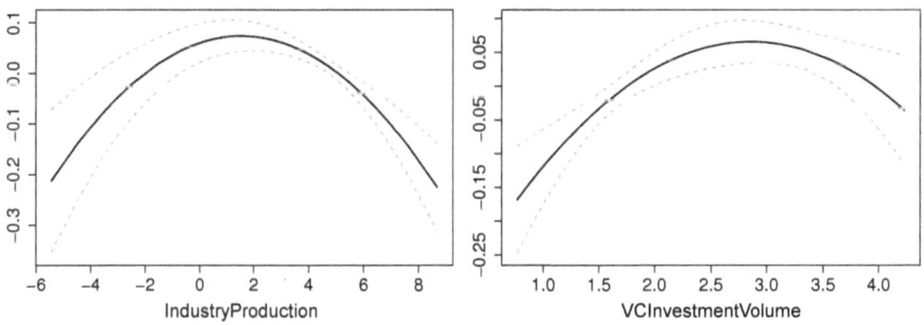

drives investment duration. I performed the analysis with the GAMLSS methodology and fitted a BCPE distribution to the data as I did for the IRR and MIRR predictions. The final result of my regressions are displayed in Table 3.9[100]; partial effects of the covariates on μ and σ are displayed in the Figures 3.10 and 3.11. The model diagnostics reveal that the BCPE distribution fits the logarithmized duration distribution very well; residuals are almost perfectly normal distributed. Therefore a Box-Cox transformation with a parameter of 3.42 is necessary. Kurtosis adaptation is also required as the statistically significant τ parameter indicates.

The regression results show that the median of the logarithm of the duration (represented by the distribution parameter μ) significantly interrelates with the return figure of an investment, i.e., very good and very bad returns occur after short investment durations while returns around 0% MIRR have longer durations. One may conclude that VC managers swiftly realize if an investment is unsuccessful. Very successful business models of portfolio companies reveal their nature quickly as well.

Industrial production growth at the time prior to an investment seems to positively influence investment duration median. A high growth rate delays

100 I tested multiple regression specifications with similar results for the functional form of the covariates. I just show the specification that has the lowest AIC/SBC. A low AIC/SBC indicates a good mix between the number of covariates and the explained deviance.

divestment possibilities, which may be explained with market cycles. The later in a growth cycle, i.e., the closer to a downturn, a fund manager invests in a portfolio company the longer it takes until the market environment recovers and the company may be sold successfully. Partial effects of the industrial production growth on σ indicate that the spread of the duration distribution pattern decreases for very positive and negative growth rates. This observation implies that fund managers act more synchronous in very good and very bad economic times. In good times they all wait longer before exiting investments and in bad times they all divest their investments sooner.

Regression results suggest that investment competition – measured by the covariate VCInvestmentVolume – seems to negatively impact the median log-duration. A strong competition at the time of an investment leads to very short durations. Probable explanation could be that fund managers tend to buy relatively more unprofitable investments in a competitive market environment. Afterwards, they soon realize that investments are unprofitable and write them off. Partial effects of investment competition on the spread of the duration distribution σ reveal that low and high competition synchronizes fund managers' behavior.

Early stage investments require more time than later stage investments before they are divested. This may be traced back to the fact that the portfolio company's business models are less developed and as such less attractive for potential buyers.

Finally, I find that during the Internet bubble fund investments were liquidated more quickly than in the rest of the time.

3.6 Conclusions

Recent publications on VC investment performance has risen doubts whether VC investing makes sense from an investors perspective or not. Researchers came to the conclusion that a high profile investment selection and fund selection as well as good investment timing are most important for success in this market. Thus, VC fund managers and VC investors require tools to limit the risks and focus on the chances. The macroeconomic environment is one important factor driving chances and risks. Thus, investors should be interested in predicting the effects of the environment on the chance-risk-profile of their investments. Existing tools for this purpose which were suggested by other authors have multiple weaknesses. In the current study I have illustrated a new methodology which compensates these weaknesses.

My methodology splits in two parts – an approach to investigate default rates and an approach to analyze non-defaulting investments. I apply two different statistical methodologies to predict the default rate of an investment. Both concepts are on the basis of credit default prediction methodologies. The first approach, based on the Credit Portfolio View (CPV), investigates and projects the default probability of individual investments with a LOGIT model. The second method is based on the Credit Risk Evaluation Model (CREM) and takes the assumption that investments in a group have the same risk profile. Thus, individual default risks are derived from the default risk in the group. Both approaches have their advantages and disadvantages; the main difference is that the CREM does not suffer from (slight) numerical downward bias but requires more observations than the CPV. Both the CPV and the CREM approaches reveal that the competition for investments is the key parameter for write-offs. Although a final conclusion is not possible, my analysis indicates that the CREM approach is slightly superior to the CPV approach concerning its predictive power.

In the second part of my approach, I look at the parameters of the performance of non-defaulting investments. So far – according to my knowledge – a comprehensive analysis of the distribution pattern of investment returns has not been performed. Existing literature focuses on the parameters of the mean performance or of the mean of the logarithm of the performance. Factors driving the spread of the returns have been suppressed so far, although they are central to understanding VC returns as mainly extreme returns define the success of VC investing. If this is not well considered in the investigation, a performance prediction is not trustworthy. This is why I suggest to use a quite new statistical methodology called GAMLSS that allows identifying parameters of the spread of returns besides modeling parameters of the average return. I applied the GAMLSS methodology and tested many distribution patterns and their fit to the two VC investment performance measures IRR and MIRR. The best fit and predictive power is reached when fitting a Box-Cox Power Exponential (BCPE) distribution to the logarithm of the IRR or MIRR.

The GAMLSS regressions reveal that factors approximating investment chances and risks, competition for investment opportunities and the conditions of exit markets drive the VC investment performance distribution patterns.

To test the quality of my methodology I compared the predictions based on my approach with the predictions based on the most closely related approach of Buchner et al. (2010). I found superior predictive power for my methodology both concerning the mean deviance of the predicted data from the real data as well as concerning the spread of the deviance. Fund managers and investors should be able to take better investment decisions based on the results they can obtain with my methodology.

4 Study 3: European Venture Capital: What can we learn from the past?

4.1 Introduction

VC is an important factor for economic growth as it targets the financing of innovative projects (Kortum and Lerner, 2000). A constant flow of new products, services and processes is necessary for any country that wants to keep up with international competition and increase the prosperity of its people. A good example are the USA who were able to convert innovative ideas into highly internationally competitive products and services through Venture Capital in the 1980's and 1990's. Good examples are companies like Microsoft, ebay or Google. Europe was trying to follow the North American example (Bottazzi and Da Rin, 2002; Aizenman and Kendall, 2008) and established a VC market as well as IPO markets in the European ecosystem over the last 20 years.

In the 1990's this concept seemed to work. Although the USA were the lead market in Internet start-ups and European start-ups mainly copied new business ideas from the US, investors earned attractive returns by financing European start-up companies. In consequence, more and more investors became interested in the asset class Venture Capital hoping to gain high returns. The strong development and growth rates in VC funding peaked in the Internet bubble in 1999/2000.

An abrupt decline in the amount of capital raised by VC funds and in the capital invested into portfolio companies by fund managers followed. Investment levels of 1999/2000 were not reached again ever since. The Internet bubble seems to be a turning point in the relatively young VC industry in Europe. The bust was a really bitter experience investors and fund managers were confronted with. The question is how market participants reacted on this evolution, how they behaved in the aftermath of the bubble and whether their actions were successful or not.

To answer these questions, this chapter investigates the historic development of the European Venture Capital industry performance before and after the Internet bubble of 1999/2000. I find a strong difference in portfolio company returns in the two time periods. While the VC industry's success was attractive before the time of the bubble, it was very low on average afterwards. Thus, VC funds raised in and after 1998 were affected most by the bad performance of their portfolio companies after 1998.

I compare the performance of European and US portfolio companies. In line with other studies (Hege et al., 2009; Gottschalg and Phalippou, 2008) I find a significant underperformance of European *early stage* investments compared to their US counterparts. However, when *overall* VC investments are considered, US investments were not statistically significant better than EU investments. This seems a consequence of the quite similar performance of European later stage investments to US later stage investments.

In the second part of this study I deal with the question if some Limited Partner (LP) groups (i.e., investor groups) were able to positively separate from the bad mean performance in Europe by superior fund selection. Although differences in the performance of LP groups become evident, they are not wide. It seems that all European fund managers had difficulties in the aftermath of the bubble. The low performance spread between the best and the worst funds naturally transforms to a quite low difference between investor groups. Corporations seem to have been the worst investors while Banks, Government agencies and GP of funds positively separated from Corporations.

Another important question relates to the reasons for the bad performance of funds raised in and after the bubble. It seems that the overall market environment was quite favorable for VC investments in the bubble aftermath. Investment conditions were good for fund managers with attractive valuations of companies especially in the years 2002-2004. The following years offered opportunities for the portfolio companies to grow, and exit market conditions were good in 2006/2007 - both concerning valuations and the possibility to sell via IPO. However, the European VC market suffered from serious capital oversupply in comparison to the available high-quality start-up companies. The difference between the cumulative capital raised and the cumulative capital invested in a year tripled between the 1990's and the bubble's aftermath. Moreover, the capital was committed procyclically by LPs and invested procyclically by fund managers into portfolio companies. This fact seems to have triggered a serious decrease of the average quality of fund managers and portfolio companies, implying a negative impact on investors' returns. Redressing the balance between capital supply and (high-quality) demand is necessary to bring European Venture Capital back to profitability for investors.

The next parts of this chapter are structured as follows: In Section 4.2 I describe findings from related literature. Section 4.3 presents the data sources I employ for this study and their representativeness. A detailed investigation of the European VC market's average performance on fund and portfolio company level is shown in Section 4.4. Section 4.5 compares the behavior and performance of different groups of investors. In Section 4.6 I investigate the reasons for the bad track record of European VC funds raised in and after the bubble, and conclusions follow in Section 4.7.

4.2 Related Literature

Several important studies have been published on the Venture Capital industry in recent years. However, summarizing, little is known about how, why and even if the VC industry works at all (Beauchamp, 2006). In this respect, my study aims to further enlighten some dark spots in the VC literature, looking critically at the industry's history.

This study may not be allocated to one specific sub-stream of the Venture Capital literature, it rather relates to different streams. The first part of my investigation is mainly concerned with the question how venture capital performed historically. Since data sources on VC investments became more and more available during the last two decades, increasing research was conducted on the net performance of VC funds and recently also on the gross performance of direct investments of fund managers into portfolio companies.

Important literature on the (net) performance of *private equity funds* includes Chen et al. (2002), Jones and Rhodes-Kropf (2003), Kaserer and Diller (2004), Kaplan and Schoar (2005), Driessen et al. (2009) and Gottschalg and Phalippou (2008). These authors investigate the performance of VC investments descriptively and in comparison to public markets. The results are not unambigous: Chen et al. (2002) analyze VC funds from the Venture Economics database between 1960 and 1999 and find an annual arithmetic average return of 45%; the geometric average return is estimated to be significantly positive at 13%.

Jones and Rhodes-Kropf (2003) show that VC investments have positive alphas[101] while investors in VC funds earn zero alphas. Fund returns increase with idiosyncratic risks; therefore, more diversified portfolios have lower average returns.

Kaplan and Schoar (2005) look at the performance of VC and Buyout funds and their persistence over time. They find significant variation in returns over time. Equally weighted PME[102] statistics show an underperformance of VC and BO funds to public markets while size weighted data

101 A positive alpha indicates an excess return over the market return on a risk adjusted basis.

102 The PME approach compares a VC investment with a public reference market. Different definitions for PME exist. Commonly, the PME measures how much one has to invest in a reference market to get the same disbursements as with the VC investment. For more details see Kaserer and Diller (2004).

unveil a slight outperformance for VC. The later result is due to the outperformance of the VC funds raised between 1988 and 1995.

In a recent study, Driessen et al. (2009) use a new statistical approach with data on VC and BO funds from the TVE database. The authors find a negative CAPM alpha of -15% for VC and a beta of 3.21.

Another recent study from Gottschalg and Phalippou (2008) who analyze a large fund dataset concludes that previous studies overstated fund performance. Their correction for sample biases changes the slight outperformance of funds to a significant underperformance of -3% p.a. vs. the S&P 500 net of fees for the period 1980-1993.

Also on the *direct investment level*, results in the literature differ profoundly about the conclusion whether or not VC outperforms public markets. Studies by Peng (2001), Cochrane (2005), Woodward and Hall (2003) and Quigley and Woodward (2003) build venture capital indices and analyze the correlation of the indices with public markets.

Peng (2001) finds an average geometric return of 55% and a high beta of 4.66 on the NASDAQ for 1987-1999.

Woodward and Hall (2003) also find an average performance of VC investments of 20% p.a. outperforming the NASDAQ with an abnormal performance of 8.5% p.a. at a beta of 0.86.

Results by Cochrane (2005) are even more bullish: He analyzed the performance of VC investments in 7,765 companies from the Venture One database correcting for sample selection and identified an annual gross return of 59% and an alpha of 32%.

These results are incompatible with Quigley and Woodward (2003) who find gross real returns of about 5% which is below S&P and NASDAQ returns in the investigated period.

Similarly, Moskowitz and Vissing-Jorgensen (2002) ascertain that "early-stage entrepreneurial investments"[103] perform below the market on a risk-adjusted basis.

103 This is a very broad definition of PE.

Ick (2005) who uses an older version of my data sample on portfolio company level finds a very mixed picture for the VC investments. Sometimes public market investments and sometimes PE investments perform better. This totally depends on the applied risk adjustment method, the benchmark used and the investigated stage or industry focusses of the investigated funds.

Overall, the results of the studies on VC performance are very mixed. I observe that the more recent the dataset is which the authors use for their investigation and the more sophisticated the authors account for distorting factors (e.g., sample selection, survivorship bias, missing data) the more often they find an underperformance compared to public markets. One reason might be that during the history of the VC market – especially in the years of the Internet bubble and its' aftermath – the industry changed significantly. The first part of my study aims to further enlighten this question. I have a more detailed look on the European VC performance over time than other authors based on my broad datasets which allow me to study both the fund as well as the portfolio company level. This allows me to identify why funds that were started at a certain moment in time performed well or badly during their maturity.

Including data on the performance of US portfolio companies, I can additionally compare European and US investments to tackle the widespread belief that European VC is worse than US VC. Other authors have also addressed this question previously: Gottschalg and Phalippou (2008) investigate a quite old dataset for the period 1980-1993 and find a worse performance of European funds in comparison to US funds, which is not surprising as the European VC market was still very young in comparison to the US during this period.

The closest study to this in respect to the US-Europe-comparison is Hege et al. (2009) who use data on portfolio company level from TVE for 1997-2003. They find that US venture capitalists generate significantly more value with their investments than their European counterparts. The authors partially explain this observation with differences in contracting behavior; other potential parameters like different law systems, stage of stock market evolution or tax subsidies in favor of venture investments seem to have no effect. Though, US trade sale outperform European sales. Hege et al. attribute the overall performance gap between both regions essentially to the poorly performing European companies. However, the datasets used in Hege et al. (2009) just includes 146 European and 233 US portfolio companies. This equals about 20 investments per year for Europe which may not be considered representative. These authors also state that their data potentially suffer from serious sample selection problems they cannot correct or adjust for. My sample base is about 7 (Europe) and 16 (US) times their size and expected not to have the same distortions.[104] Moreover, my detailed information on cashflows enables me to calculate precise return measures which was not possible for Hege et al. (2009). Thus, I enjoy to have a much more detailed look at the returns in the two markets and their relation to each other.

The second part of my investigation is related to another set of studies on the fund level which analyze how and how successfully different investors (often also referred to as limited partners (LPs)) invested in VC funds. A comparison study which investigates the sources of capital from VC funds in Germany, Israel, Japan and the United Kingdom is from Mayer et al. (2005). The authors analyze data from national VC associations and a survey from Financial Nikkei in Japan. Mayer et al. find that banks are a major VC investor in Germany and Japan while Israel heavily relies on Corporations and the UK on Pension funds. VC funds that are funded by Banks and Pension funds invest more in later stage activities than funds that source their capital from Individuals and Corporations. However, the sources of funds may just explain a small portion of the variation in the funds' investment activities.

104 For the period after 1998 I have 4-5 (Europe) resp. 8-9 (US) times their amount of data.

The first comprehensive study on investment success of different investor classes is from Lerner et al. (2007) who use a quite old dataset to investigate the time-frame 1991-1998 for the USA. The authors find a performance of 21% on average for endowments which is superior to all other LP types. They conclude that endowments are better in predicting if a follow-on fund will be successful. The superior returns are not due to better connections of the endowments to established funds. Banks and corporate pension funds were the worst investors in the study.

Two working papers from Hobohm (2008a,b) look at Prequin data and extend the LP investment success analysis of Lerner et al. (2007) time-wise (investigating data between 1991 and 2003), geographically (with a world-wide focus) and stage wise also looking at buyout funds. Hobohm (2008b) finds that investor success differs significantly for different stage focuses. Endowments, investment corporations and insurances seem to be good PE investors while Government agencies were worst. Though, this result seems to be very time dependent and equalling out in the long run. In the period after 1998 the performance differences between LPs became significantly lower than in the time before. The author states that proximity matters for PE investing, and closeness between LPs and VC funds positively affects investment returns for LPs. In his second study, Hobohm (2008a) also identifies a "Home bias", i.e., 81% of investments by an average LP are put into funds on the same continent. Moreover, the study shows a strong relation between the size of an LP and the size of a fund. Unsurprisingly, bigger LPs prefer bigger funds to invest in, e.g., due to limited resources for screening funds. In respect to the success of endowments, Hobohm's results are quite contradictory to the study of Lerner et al. (2007). Hobohm finds that endowments were investing successfully up to 1998 (due to very good early stage fund picking) but performed among the worst afterwards. For Banks and Investment Corporations the results are inverted. They performed better after the Internet bubble than before.

The studies of Lerner et al. (2007) and Hobohm (2008b,a) do not focus on European VC exclusively. Thus, their results may be skewed by observations from buyout funds or international funds. My investigation on LP performance may be seen as complementary – especially to the results from Hobohm (2008b) – as I am able to analyze exclusively the European VC market in the bubble and the aftermath with a new dataset.

The third part of my study investigates the reasons for the low performance of European VC in the bubble and its' aftermath. It is related to articles on VC performance parameters (Gompers and Lerner, 2000; Ljungqvist and Richardson, 2003; Kaplan and Schoar, 2005; Kaserer and Diller, 2004; Gottschalg et al., 2004). In their "milestone" study, Gompers and Lerner (2000) highlight the negative impact of high inflows of capital into VC funds on their performance ("money chasing deals" phenomenon).

Kaplan and Schoar (2005) report that more experienced fund managers have a persistently better performance. Better performing fund managers have a higher chance to raise a follow-on fund and new funds are more negatively affected by the market cycle than established funds.

Ljungqvist and Richardson (2003) notice that the TVE data used by other authors suffer from biases due to self-reporting and accounting problems. Their cash flow data is from one of the largest US PE investors and seems to be more appropriate to analyze performance. The authors show that fund managers react on changes in competitive conditions in the PE market – mainly concerning competition for deal flow – when timing investments and exits. Their results are central to understanding parameters of capital drawdowns and capital disbursements within funds.

In a study including 200 mature European funds between 1980 and 2003 from the TVE database Kaserer and Diller (2004) find that – apart from risks – the PE market capital flows, market sentiment and skills of general partners affect fund performance. Gottschalg et al. (2004) show that fund performance positively co-varies with business and stock market cycles; returns are highly depending on the valuation levels at the time when investments are exited. Nowak et al. (2004) state that investment timing is more important than exit timing for VC funds.

My investigation on reasons for the bad performance of the European market after 1998 also relates to studies on VC market structures. For example, Meyer (2007) and Gilson (2003) investigate how a VC market may be built focusing on the role that governmental institutions can and should play. Gilson (2003) emphasizes that many mistakes have been made internationally and especially in Europe in trying to copy the US VC market model. But Europe learned from these mistakes in the last decade. Another study of Meyer (2008) analyzes the effect of innovation financing through VC on economic growth; he suggests that Europe doesn't miss capital but entrepreneurs.

However, none of the above mentioned studies has a comprehensive look at the multiple dimensions that led to the recent problems of the European VC market. This study aims to close this gap.

4.3 Data Description

Before going into detail on my analysis, this section presents the investigated datasets and deciphers their representativeness. I use mainly three data sources:

1. Information on European Venture Capital funds was provided by institutional investors in Europe who want to stay anonymous. The samples I received contain 176 VC funds that were raised between 1994 and 2008 with the majority of funds raised after 1997. The advantage of my fund sample is that I have detailed information on the funds themselves including the funds' net cashflows (i.e., after management fees and carried interests), the funds' investments into portfolio companies as well as the composition of their LPs.
2. My second dataset was provided by the *CEPRES*.[105] This company kindly gave me access to detailed information on VC financed European portfolio companies and their cashflows. *CEPRES* is a subsidiary of the Sal. Oppenheim Bank which collects detailed information on

105 I especially thank Dr. Daniel Schmidt and Dr. Philipp Krohmer for providing the datasets on US and European portfolio companies.

24,062 investments of 1,103 worldwide VC/PE and Mezzanine funds (status as of September 2009). The data are totally anonymous and very accurate; they are directly reported by the GPs and not collected through secondary research like other commercially available sources (e.g., *VentureXpert* or *Prequin*). The *CEPRES* data are collected through standardized information request sheets and validated with due diligence reports, including audited filings, to ensure accuracy.

CEPRES data have been used by other authors before, e.g., Nowak et al. (2004), Cumming and Walz (2004), Krohmer and Man (2008), Ick (2005), but have grown since then. My *CEPRES* sample for Europe consists of 1,111 investments in 898 companies between 1989 and 2007 from 188 funds of 94 fund managers.

The portfolio companies are all either fully or partially exited or written off. This allows me to rule out distortions from subjective valuations and thus implies a reliable measure for their performance. Especially, the problem of a survivorship bias is minimized (i.e., my sample is expected to contain all successful investments as well as failures of the VC funds). The cashflow information from *CEPRES* is gross, i.e., before management fees and carried interest payments to the fund managers.

For comparing the performance with the US VC market, I also received a data sample from CEPRES consisting of 4,046 fully and partially realized or written-off VC investments in the US. I do not go into detail on the composition of the US sample at this point. A description of the fully realized and written-off investments can be found in Section 3.4.

The detailed cashflow data on fund **and** portfolio company level allows me to decipher the investment behavior of fund managers and have a detailed look at the profitability of investments in different economies, industry sectors, stages as well as over time.

3. I also obtained data on 1,756 European VC funds from *VentureXpert* to check the representativeness of my fund sample and perform some more rough investigations.

Supplementary data for my analysis are from the *European Venture Capital Association (EVCA)*, the *British Venture Capital Association (BVCA)*, *Eurostat*, *Thomson Datastream* and *Thomson One Banker*.

4.3.1 The Fund Sample

4.3.1.1 Composition

The overall fund sample contains 176 VC funds raised between 1994 and 2009 in total. A fund requires at least 3-4 years to start the main share of its' investments. So I restrict the funds used for my analysis to the funds raised 1994-2005 (called *full fund sample* in the following), i.e., my funds are at least four years old. As I require quite mature funds for some parts of the investigation (to mitigate distortions from the subjective valuation of non-liquidated portfolio companies of the funds) I create a sub-sample (called *restricted fund sample*) which just contains funds that are older than seven years. An analysis of Kaserer and Diller (2004) shows that the IRR of a fund four years before liquidation is a fairly good estimate for the final performance.[106] Thus, my full fund sample contains 138 funds and the restricted fund sample 109 funds.

The distribution of the funds in the full fund sample across countries is illustrated in Figure 4.1. The dark gray bars show in which country a fund has his main office. Most of the funds are located in the biggest European economies UK, France and Germany. This is due to the historic development of the VC industry in Europe. Interestingly, the funds in the three main European economies like to focus on other countries for their investments or do not specify any country focus (called Multi-Country) more often than the funds of the smaller economies. This reflects the evolution of the industry in

106 The average difference between the IRR four years before the fund reaches its' maturity and the final IRR is just a few percentage points in their analysis.

Figure 4.1: Distribution of full fund sample across countries (Source: Full fund sample)

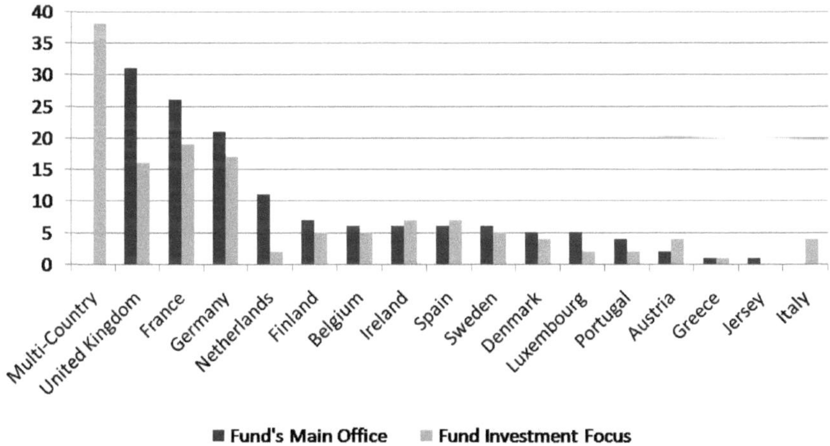

the respective countries. Going international is usually a second step after investing nationally (Aizenman and Kendall, 2008).[107]

Table 4.1: Focus of funds in full fund sample along stages and sectors
The Table shows the distribution of the number of funds raised by vintage year and their respective stage and sector allocations.

Vintage Year	1994-1997	1998	1999	2000	2001	2002	2003-2005	Total	in %
Stage									
Early Stage	5	12	10	23	22	15	9	96	70%
Later Stage	1	1	6	1	5	3	2	19	14%
Balanced VC	1	4	2	2	4	2	8	23	17%
Sector									
Generalist (Non-Tech)	0	2	3	1	3	2	2	13	9%
ICT	1	7	5	11	13	7	8	52	38%
ICT-Life Science	6	8	7	11	9	7	6	54	39%
Life Science	0	0	3	3	6	4	3	19	14%
Total	7	17	18	26	31	20	19	138	100%

Table 4.1 shows the focus of the funds in the full fund sample along stages and sectors over time. 70% of all funds in the sample are early stage funds. The dominant sector focus is ICT (Internet, Computer and Telecommunication) sometimes combined with Life Science. Only 9% of the funds lack a focus. The historic development in the sample is due to

107 For a similar observation see Meyer et al. (2005, p. 593).

the development of the investors who provided the data. Considering the distribution of the funds in the full fund sample over time I can see that the investors were mainly investing in the vintage years 1997 until 2002.

4.3.1.2 Representativeness

The fund sample contains information of few investors in Europe. This naturally raises the question how representative the sample is and if the results obtained in my investigation may be generalized to the European market.

Therefore, I first assess the share of the fundraising volume in the investigated time-frame contained in my sample. This analysis requires data on the total number and amount of funds raised in Europe in the investigated period. Two institutions collect the information: EVCA and VentureXpert. From EVCA I received fundraising data for 1998 and thereafter, the VentureXpert database contains data since 1994. Both EVCA and VentureXpert collect data on the volume raised in *a certain year*, i.e., if a fund management team takes two years to raise sufficient capital to close their fund, the fundraising volume of this specific fund is split between the two years. Moreover, one has to keep in mind, that the EVCA sample contains the volume which is *expected* to be invested in VC in Europe. This means, the numbers contain fund volumes of Buyout funds which are expected to be invested in venture companies.

On the contrary to the EVCA and VentureXpert data, my fund sample just shows the vintage year of a fund and does not allow to differentiate in which year the capital has been raised. Thus a direct comparison of the different datasets is not possible. To allow a conservative estimate, I compare the fund data of 1999-2005 with the EVCA and VentureXpert data from 1998-2005, i.e., I add the fund volume of one fundraising year in the VentureXpert and EVCA data. Table 4.2 shows the comparison. My fund sample has a coverage of 11.2% of the VentureXpert volume and around 8.8% of the EVCA volume which seems to be a good sample. The Ventur-

Table 4.2: Comparison of committed capital volumes in full fund sample with EVCA data and VentureXpert data

Fund Year or Vintage Year	EVCA (expected allocation to VC, by Year)	VentureXpert (by Year)		Full fund sample (by Vintage Year)	
	Committed capital (EUR Mil)	No. of Funds	Committed capital (EUR Mil)	No. of Funds	Committed capital (EUR Mil)
1994-1997	-	273	10,639.70	7	451.56
1998	6,730.78	156	5,600.70	17	1,223.19
1999	12,091.74	243	10,158.80	18	1,037.54
2000	22,144.93	444	26,622.80	26	1,751.32
2001	15,029.06	223	9,294.20	31	2,026.76
2002	8,528.87	109	3,159.30	20	839.60
2003-2005	25,376.23	295	15,451.00	19	2,242.57
as of VentureXpert 1998-2005					11.2%
as of EVCA 1998-2005			78.2%		8.8%

eXpert volume accounts for 78% of the EVCA volume in the time-frame. As VentureXpert does not allocate volumes because of expected allocations to VC (which is arbitrary), but just uses the fund stage focus, I consider these numbers as the point of reference for further investigations.

With approximately 10% of the total fundraising volume represented in the full fund sample it is required to assess potential selection biases in the data. Therefore, I compare my full fund sample with VentureXpert. I know all fund names for the fund sample and I can determine if a certain fund is also included in the VentureXpert database. For 92% of the funds in the full fund sample I find a match in the VentureXpert database. This allows me to detect parameters that show me if a VentureXpert fund is also in my sample. First, I compare the two samples concerning fund sizes and risk structure in a descriptive analysis.

Fund size is often seen as an indicator for experience and more experience should lead to better performance (e.g., Kaplan and Schoar, 2005). Using a comparison of the fund size distributions in both data samples, I want to answer the question if the fund sample contains more or less experienced fund managers. An untabulated Mann-Whitney-Wilcoxon test[108] shows that my

108 The Mann-Whitney-Wilcoxon test is a rank sum test. Its field of application is quite similar to a t-test, but does not require normal distribution of the variables which are compared.

sample is biased towards larger funds (p-value: 0.00***), i.e., I expect to have rather experienced fund managers in my sample.

Table 4.3: Comparison of stage focus of funds in full fund sample and in VentureXpert (Please note that balanced stage funds are split 50:50 on early and later stage focus for the full fund sample.)

	Early stage	Later stage
VentureXpert	1318	438
% of total	*75%*	*25%*
Full fund sample	107.5	30.5
% of total	*78%*	*22%*

The stage focus of a fund is considered as an indicator for risk (e.g., Kaserer and Diller, 2007; Buchner et al., 2010; Krohmer and Man, 2008). Along the argumentation of the *Capital Asset Pricing Model*, risk should be positively correlated with returns. Thus, if my sample is biased towards more risky investment strategies this should theoretically result in higher returns. Table 4.3 shows the distribution of the funds' stage focus in the two samples. Although a clear bias to early stage focused funds cannot be identified, it might be possible.

A clearer picture may be obtained through regression analysis. Table 4.4 shows the regression results. I calculate several specifications of LOGIT models on the VentureXpert fund data. The dependent variable is a dummy which indicates if a fund is in my fund sample (dummy = 1) or not (dummy = 0). I use the fund size, the number of portfolio companies of the funds and the funds' stage focuses as explanatory variables.

Obviously I am facing a bias towards larger funds. The influence is statistically significant both in the fund size specification and in the joint specification. The second tested indicator for size and for investment activity is the number of portfolio companies of a fund. This variable is highly statistically significant in all specifications as well pointing to more active funds in my sample. The negative and statistically significant later stage indicator reveals that the fund sample may be biased towards early stage focused funds.

Table 4.4: Logistic regression to detect potential biases in full fund sample
The presented analysis is based on VentureXpert data from the years 1994 to 2005. I calculate four specifications of Logit regression models with a dummy as dependent variable. This dummy indicates if a fund from the VentureXpert sample is also part of my full fund sample. The independent variables are the logarithmized fund size in Euro, the number of investments of the respective fund and a dummy variable which indicates if a fund focusses on later stage investments (=1) or early stage investments (=0). Standard errors are in brackets. ***, **, * indicate statistical significance at the 1%, 5% and 10% significance level. Pseudo-R^2 is Nagelkerke's R^2.

Logistic Regression on Dummy Variable (1 = VentureXpert fund is in full fund sample; 0 = otherwise)				
	Stage Specification	Fund Size Specification	Number of companies Specification	Joint effects Specification
(Intercept)	-1.980***	-3.697***	-3.445***	-3.691***
	(0.615)	(0.699)	(0.665)	(0.703)
Later stage	-0.708***			-0.860***
	(0.272)			(0.289)
Fund Size (log)		0.536***		0.440***
		(0.086)		(0.096)
Number of Companies (log)			0.738***	0.418***
			(0.103)	(0.122)
Vintage Year Fixed Effects	Yes	Yes	Yes	Yes
Observations	1,537	1,054	1,534	1,051
Pseudo-R^2	4.7%	12.0%	12.1%	16.1%

4.3.2 Composition and Representativeness of the CEPRES Sample

The structure of my second European sample, the one from *CEPRES*, is illustrated in Table 4.5. It contains 1,111 investments over the time period 1989-2007. 80% of the investments are fully realized, i.e., for 20% of the investments there might be some bias from the valuation of remaining shares in the portfolio of the reporting fund. I expect this bias to be minor because the valuation is not based on pure estimations; instead, a price of the portfolio company was already identified during a partial sale.

The share of early stage VC investments in the sample is about 48%, 34% of the investments are in later stage companies, the rest is unspecified in the *CEPRES* database. The relative importance of the stages changed significantly over time. While early stage investments were very important in the beginning and the end of the 1990's, later stage dominated between 1994 and 1997. This result is not surprising as anecdotal evidence suggests that in the advance to the Internet bubble more and more early stage IT companies were financed. This is confirmed by the distribution of investments across industry sectors. As the table shows, in 1995-1997 the main share of investments was in the Industrials sector while the Computer sector dominated in 1999 and 2000.

A complete assessment of the *CEPRES* sample's representativeness is not possible as I lack a reference sample to compare with. However, I have no reason to believe that the *CEPRES* sample suffers from any bias based on my descriptive analysis. The sample has a large size and both defaulted investments as well as investments successfully sold via M&A and IPO are included in the sample.

4.4 Historical Performance of the European Venture Capital Industry

4.4.1 Performance Measures Applied in this Study

Measuring the performance of VC investments and funds is not as simple as in public markets. The main reasons are the illiquidity and the limited transparency of the market as well as irregular observations of valuations. Especially the fact, that actors in the private equity market hardly publish returns and that the VC industry lacks a well functioning secondary market, leads to infrequent and publicly unobservable investment successes and failures. Even researchers able to build on very detailed information of funds and investments into portfolio companies may just approximate a fund's performance development. Moreover, measuring performance is not as simple as in public markets, where the market price is assumed to contain all information that is publicly available and researchers just have to calculate a return development, based on the price development.

I calculate return figures both for single portfolio companies and for whole VC funds in this study. The same fundamental principles are applied in both

Table 4.5: Composition of CEPRES sample of VC investments along realization status, stage and industry
Note: Due to confidentiality reasons I do not show all precise numbers of investments.

Investment Year	**1989**	**1990**	**1991**	**1992**	**1993**	**1994**	**1995**	**1996**	**1997**	**1998**	**1999**	**2000**	**2001**	**2002**	**2003**	**2004**	**2005**	**2006**	**2007**	**Total**
by Realization status																				
Fully realized	<5	15	27	20	25	44	41	50	70	84	127	237	76	24	13	13	13	<5	<5	884
Partially realized	<5	<5	<5	<5	<5	<10	6	<5	20	23	28	36	34	19	14	13	13	5	<5	227
by Stage																				
Early stage	<5	7	11	14	15	12	19	17	21	55	85	169	54	19	7	11	<10	<5	<5	528
Later stage	<5	5	<10	5	8	30	21	28	68	42	40	61	30	12	10	7	<5	<5	<5	382
Unspecified VC	<5	<5	10	<5	<5	<10	7	9	<5	10	30	43	26	12	10	8	14	<5	<5	201
by Industry																				
Computer	<5	<5	<5	7	<5	8	12	9	9	27	86	158	57	19	14	12	11	<5	<5	441
Consumer	<5	<5	<5	<5	<5	6	<5	<5	7	6	<5	<5	<5	<5	<5	<5	<5	<5	<5	45
Healthcare/LS	<5	5	7	6	7	8	8	9	11	27	22	23	18	7	<5	<5	5	<5	<5	170
Industrials	<5	5	8	<5	9	17	17	26	51	28	19	36	15	7	7	9	<5	<5	<5	267
Telecom	<5	<5	<5	<5	<5	6	<5	5	8	8	19	44	15	5	<5	<5	<5	<5	<5	126
Others/Unspecified	<5	<5	6	<5	<5	<5	5	<5	<5	11	7	8	<5	<5	<5	<5	<5	<5	<5	62
Total	**<5**	**<20**	**<30**	**<30**	**<30**	**<50**	**47**	**54**	**90**	**107**	**155**	**273**	**110**	**43**	**27**	**26**	**26**	**<10**	**<5**	**1,111**

cases. Though, one has to keep in mind that fund returns are net of fees and carried interest payments while portfolio company returns are gross.

To approximate VC returns, the most common measure in the industry are multiples and the IRR. I calculate the multiple "Total Value to Paid-In (TVPI)" which equals the ratio of the sum of all disbursements and the net asset value as of the sum of all investments. The IRR of an investment is defined as the average return of the invested capital over the investment period, i.e., it is calculated by solving

$$\sum_{t=0}^{T} \frac{C_t}{(1+IRR)^t} + \frac{NAV}{(1+IRR)^T} = 0 \quad \text{for } IRR.$$

Here, C_t are the (positive or negative) cashflows in period t, with T standing for the investment maturity. NAV is the net asset value of the unrealized and/or partially realized investments. The higher the share of portfolio companies of a fund is, which are not yet fully exited, the more subjective the performance measure is. To compensate for this problem, I just use the restricted fund sample in the investigation. Subjectivity in the analysis on the *CEPRES* data is limited as explained before.

Beyond the IRR, which is an absolute performance measure, I benchmark the VC performance with the performance of the European public stock market indices MSCI Europe and Dow Jones EuroStoxx Small Cap. I chose these two indices because the MSCI Europe reflects the stock market development of the major European economies and the Dow Jones EuroStoxx Small Cap seems applicable to approximate the development of young and more risky companies like VC portfolio companies. The advantage of a comparison to public markets is that I take general market developments into account. In periods of badly performing stock markets VC investments may suffer as well. In a comparison, I can uncover whether public or private markets performed better.

To measure the relative performance of VC I calculate Excess-IRRs and PMEs based on the methodologies of Ick (2005) and Kaplan and Schoar (2005). The first step to determine Excess-IRRs is to simulate cashflows for investments into the stock indices that are similar to the VC cashflows (of funds or portfolio companies depending on which level is investigated). Naturally, the last simulated cashflow in a cashflow series usually diverges from the last VC cashflow which results in a different return (for details see Ick, 2005). To calculate Excess-IRRs I subtract the IRR of the simulated cashflows from the IRR of the VC fund or VC investment.[109] A positive Excess-IRR means VC outperformed the benchmark, a negative Excess-IRR states the public market performed better.

The PME equals the Euro amount one would have to invest in a chosen benchmark to generate a return equal to that of a one Euro investment in VC on a present value basis (Ick, 2005, p.12). The measure implicitly takes the assumption that intermediate cashflows are reinvested in the public market index and determines the value of the cashflows accordingly. Different definitions for the PME exist in the literature. I calculate the PME as the ratio of the discounted value of all cash outflows (disbursements) and the discounted value of all cash inflows (investments). The discount rate equals the return of the benchmark index in the time-frame of the VC investment. A PME greater than one indicates outperformance of VC versus the public benchmark, a PME of less than one means underperformance.

A single return measure may distort the real picture (Askar, 2008) as all the measures in VC have strengths and weaknesses and may show different results as they all depend on specific assumptions. My approach to use several measures should, however, compensate for this problem.

109 In the cashflow simulation, I use an approach that avoids short positions in the public market.

4.4.2 Development of Venture Capital Before and After the Internet Bubble in Europe and the US

Prior to performing some detailed investigations on the performance of European VC I roughly look at the development of fund IRRs over time comparing European and US fund returns based on *VentureXpert* data. Anecdotal evidence suggests that European VC funds performed significantly below their US counterparts.

Figure 4.2 seems to confirm the significant difference in performance between European and US Venture Capital funds in the middle of the 1990's based on *VentureXpert* data – which we, however, cannot confirm with *CEPRES* data in Section 4.4.5. US funds of the vintage years 1994-1997 reached Pooled-IRRs[110] and Capital Weighted Average IRRs of 40% p.a. and more. European funds did not surpass the 25% p.a. IRR since inception. This difference probably originates to the maturation of the different market environments. While the US VC market has a long history back to the 1960's and before, the European VC market mainly developed in the late 1980's. Moreover, the US stock market (especially the National Association of Securities Dealers Automated Quotations (NASDAQ)) which is a key success factor for the US Venture Capital industry used to be a more developed exit channel than European IPO markets. A stock market environment to float shares of young (technology) companies was not established in Europe before the late 1990's (Hege et al., 2009).

In contrast to the disparity between Europe and the US until 1998, Figure 4.2 shows that funds on both sides of the Atlantic performed almost equally bad after 1998. The pooled average and the capital weighted average IRRs fluctuate around 0% in both regions. At least for the US market, there seems to be a kink in the performance curve, i.e., the levels of the 1990's were not reached again after 1998. The fluctuation in returns is also lower

110 For the calculation of pooled-IRRs the cashflows of all funds started in a certain vintage year are summed up over time as if they were all pooled in a big Fund-of-funds. The pooled-IRR is the IRR of the simulated Fund-of-funds.

after 1998. Thus, it seems that VC history may be split into a period before the bubble and a period in and after the bubble.[111] I use the year 1998 as the year of the turning point.

4.4.3 Net Performance of European Venture Capital Funds

I start my detailed analysis of the performance of European VC funds investigating the development of the interim IRRs in the full fund sample over time. Therefore I split the sample by vintage year. I expect a J-curve pattern of the returns of funds started in the same year. The average fund return of a young VC fund should start with a negative value due to the investments in the beginning of a fund's lifetime, recover over time and reach a positive level at the end of the maturity (usually after 10 years). Figure 4.3 shows the real development of the funds with different vintage years (top chart) and the parallel development of the MSCI Europe (bottom chart). In order to come up with the Interim-IRR curves I regressed Interim-IRR

111 Meyer (2008) comes to the same conclusion.

Figure 4.2: Comparison of IRRs of European funds and US funds with vintage years 1994-2005 (Source: VentureXpert)

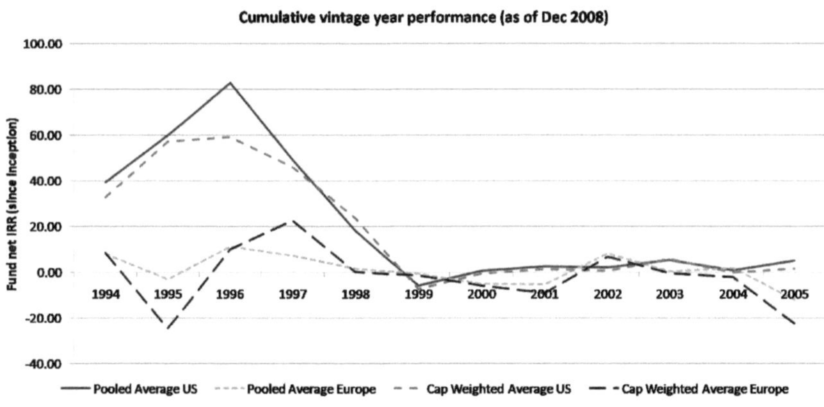

observations of the sample funds on the time it took to get the Interim-IRR observations by fitting splines with four degrees of freedom.[112]

Figure 4.3: Development of Interim-IRRs of European Venture Capital funds of the vintage years 1997-2005 over time (Source: Full fund sample)
The chart on the top shows Interim-IRR curves for different fund vintage years. The result were obtained by regressing Interim-IRR observations of the sample funds on the time it took to get the Interim-IRR observations by fitting penalized splines with four degrees of freedom. The bottom chart shows the development of the MSCI Europe index between 1997 and 2009.

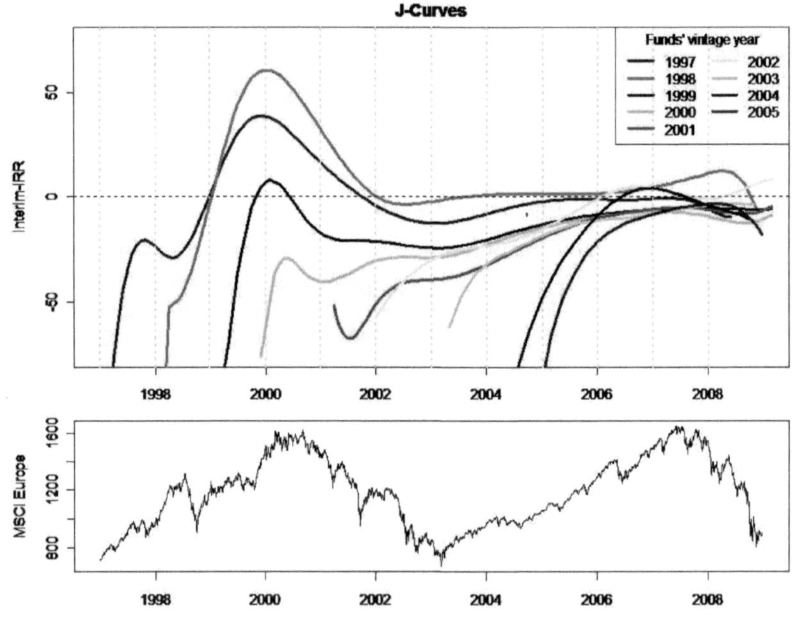

Interestingly, I do not find the typical J-curve pattern for the vintage years 1997-2000. The stock market situation in 1999 and 2000 rather gave the fund managers exit opportunities so exceptionally favorable that fund IRRs were pushed into positive territory within a very short period of time. A similar effect, i.e., a quick rise in Interim-IRRs, also appeared with the funds of the vintage years 2004/2005 in 2007/2008. However, the effect at those times was much lower although the MSCI Europe reached the same

112 I applied penalized smoothing splines: Therefore, the degrees of freedom are penalized in the fit leading to a smooth spline with relatively few degrees of freedom that well approximate the data. For more details see Eilers and Marx (1996).

high levels again in these few years after the Internet bubble. This observation suggests that the investment behavior has changed in the interim time. Either fund managers did not bet on the right sectors that picked up between 2005 and 2007 or they invested too late in the recovery. I will investigate this aspect in more detail in Section 4.6.

The negative development of Interim-IRRs between 2001 and 2003 is a result of the retreating stock markets and points to unsuccessful investments during the time of the Internet bubble. The decrease in IRRs in the years 2007/2008 is much lower than in 2001-2003. This indicates a generally low performance in the after-bubble-period which is confirmed by the fact that most Interim-IRR curves of the different vintage years end up in negative territory in 2009.

Figure 4.4 summarizes the mean and median performance of the VC funds in the restricted fund sample using the performance measures introduced in section 4.4.1. I use the restricted fund sample with more mature funds for this investigation as the returns of younger funds may be skewed due to subjective valuations of the high share of Net Asset Value (NAV)s in the funds.

The six charts in the Figure 4.4 illustrate a serious problem for the European VC industry. All measures indicate bad returns of the funds. The weighted average (median) IRR - which takes the different fund sizes into account - is -7% (-8%) for early stage focused funds and -9% (-16%) for later stage focused funds. The overall weighted average and weighted median lie at -7% and -10%. Fund managers lost 22% of their capital (on the weighted average) and 35% on the weighted median as the TVPI chart shows.

One could argue that the fund managers were facing terrible conditions especially in the beginning of the new millennium and stock markets did not perform well, too. The comparison of the fund performance with public benchmarks allows to test this conjecture. In fact, compared to the (broad benchmark) MSCI Europe, the Excess-IRR shows that the VC fund underperformance in percent is less negative than the absolute performance figure.

Figure 4.4: Performance of European Venture Capital funds from the restricted fund sample measured with Net IRRs, Total Value to Paid-In, Excess-IRRs vs. the MSCI Europe and the Dow Jones EuroStoxx as well as Public Market Equivalents (PME) vs. the MSCI Europe and the Dow Jones EuroStoxx

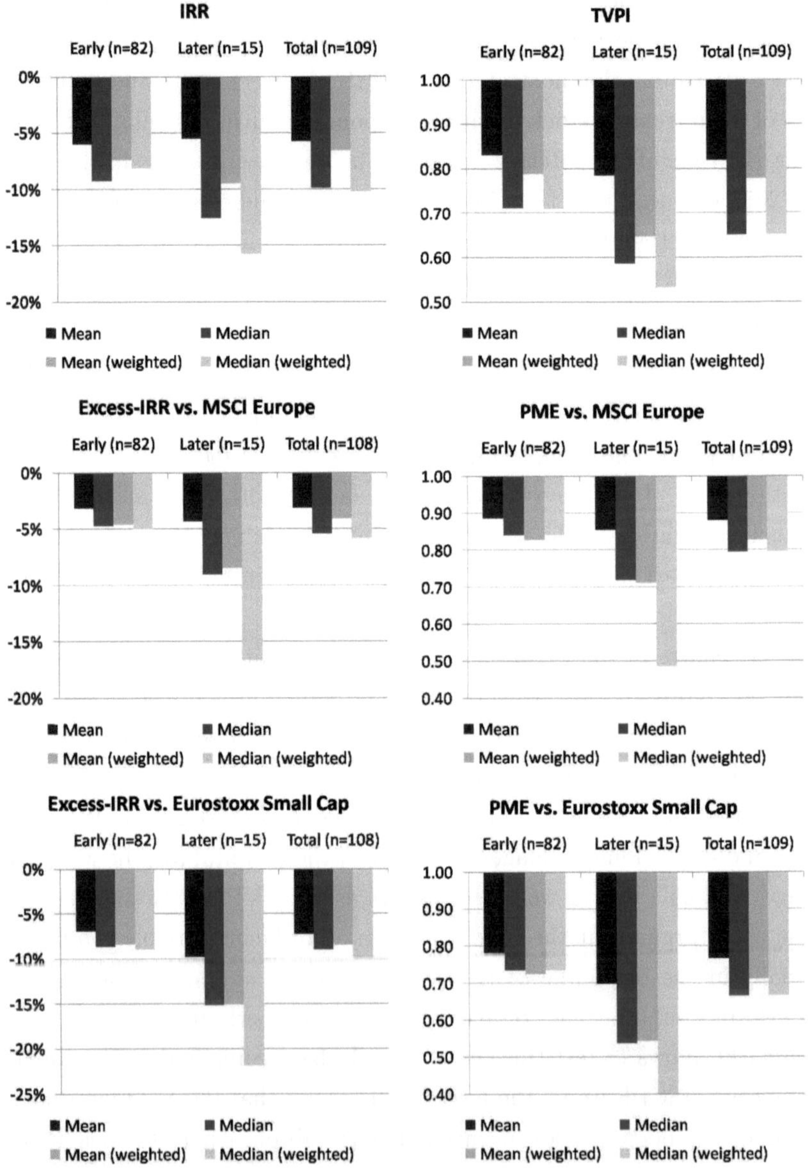

However, the funds in my sample still show a negative overall picture: The weighted average PME vs. the MSCI Europe is at 0.83 which confirms a strong underperformance. The comparison with the DJ EuroStoxx Small Cap reveals even worse relative returns. This benchmark is closer to VC than the MSCI Europe in the risk structure. The annual Excess-IRR is at -8% for early stage focused funds and -15% for later stage focused funds (weighted average). Thus, investors would have been better off, if they spent their capital on public equity markets than on VC funds.

4.4.4 Confirmation of Negative Fund Performance by Other Studies

My findings on the performance of European VC funds mainly consider the time frame 1998 and thereafter. In section 4.3.1.2 I assessed the representativeness of the sample and found some potential selection biases. This is why I investigate if my findings are specific to my sample from the anonymous data provider or a real problem in the industry. Therefore, I use industry statistics from other sources for comparison with my results.

Table 4.6: Comparison of fund performance data from VentureXpert, BVCA and my full fund sample (Sources: VentureXpert database, BVCA Performance Measurement Survey 2008, Full fund sample)

Vintage Year	Net IRR since inception by vintage year					
	VentureXpert (as of December 2008)		BVCA (as of December 2008)		Full fund sample (as of August 2009)	
	n	Mean	n	Mean	n	Mean
1994	19	6.7%	NA	NA	1	NA
1995	21	1.1%	NA	NA	0	NA
1996	21	13.0%	1	NA	1	NA
1997	44	8.1%	10	21.3%	5	-1.0%
1998	49	4.2%	1	NA	16	6.2%
1999	75	3.8%	12	-6.3%	17	-7.1%
2000	116	-4.0%	13	-6.9%	26	-12.4%
2001	69	-3.7%	12	-5.1%	30	-10.0%
2002	38	0.2%	14	6.0%	20	-3.4%
2003	36	-1.7%	8	-4.1%	2	-5.2%
2004	38	-5.8%	2	NA	2	-17.2%
2005	27	-12.3%	NA	NA	15	-5.9%
Statistics as of 1998-2005						
Mean		-2.4%		-3.3%		-6.6%
Weighted Mean		-1.9%		-2.8%		-6.7%
Share of funds raised	30%		4.2%		8.7%	

Table 4.6 shows a comparison between the full fund sample[113] and industry statistics from VentureXpert (which were also the basis for Figure 4.2) and from a BVCA study (BVCA, 2008). In the relevant time-frame 1998-2005 the performance in the industry statistics shows a negative weighted average (weighted by the number of funds in the vintage years) of -1.9% and -2.8% for the VentureXpert and the BVCA data. Thus, both are also negative. The corresponding figure in my fund sample is -6.7%. Although the 4-5% difference between these averages might indicate a downward bias in my data on the first sight, I believe this bias is only minor, if not negligible. This thesis is based on three facts:

- The industry benchmarks from VentureXpert and BVCA are both showing the status as of December 2008 (more recent statistics were not available). This means that effects from the financial crisis that started to become serious after the collapse of Lehman Brothers in the end of 2008 are hardly reflected in the benchmark data. However, my fund sample also reflects changes in the first quarter 2009. According to my interim-IRR analysis, this effect is -1% to -2% for my data – mainly for the years 2002 and later on. Falling stock markets and downgrades of economic perspectives affected NAVs of unrealized and partially realized portfolio companies negatively which lead to lower return figures of the younger funds in my sample.
- Other authors have found an upward bias in the (Thomson) VentureXpert data: Phalippou and Gottschalg (2009) compare VentureXpert fund performance data with VentureXpert data for which the performance is unknown. The authors identify an upward bias of at least 1%. Moreover, they state that the funds for which Thomson lacks data completely, are probably without successful exits (otherwise Thomson would have spotted them). They conclude another significant upward bias for which the extent cannot be specified explicitly. In another study, Woodward (2009) compares US VentureXpert data with Cam-

113 For three funds in the full fund sample I lack data on their historic performance. This is why I show summary statistics for just 135 instead of 138 funds.

bridge Associates and Sand Hill data. She finds that VentureXpert data has a higher alpha (indicator for outperformance) than the other sources. This points to an upward bias in the data as well.

- As I identified in Section 4.3.1.2 my sample probably contains funds from more experienced fund managers. To further validate this I compared the list of funds in the full fund sample with a list of funds that belong to management firms which are supposed to be the most successful firms in Europe (Source: PeNews August 2009). I found that my sample contains many of these firms' funds. Thus, I do not expect to have a negative selection of unsuccessful funds in my sample.

Summarizing, the sample seems quite representative concerning fund performance. If there were a bias, it is expected to be low. Further analysis based on the fund sample may at least be seen as a good indicator for the overall industry.

Thus, the average performance of the VC funds raised in Europe was negative in the bubble period and the aftermath of the bubble. Fund managers burned a lot of capital.

4.4.5 Gross Performance of European Venture Capital Portfolio Companies

General Performance Analysis In a second step for the analysis of the European VC industry performance, I analyze the fund investments into European portfolio companies. This allows me to detect differences in the performance of investments between stages, industry sectors and countries. Moreover, I can evaluate if European investments performed worse than their US counterparts over time.

Tables 4.7 and 4.8 compare the performance of investments in European and US portfolio companies along the four return measures presented in section 4.4.1. The gross IRRs since inception, presented in Table 4.7 Panel A, show an overall negative performance of European early stage investments with a mean of -10% and a median of -41%. Very few big investments pull

Table 4.7: Absolute performance of fully and partially realized European and US Venture Capital investments into portfolio companies measured in Gross IRR since inception and Total Value to Paid-In (TVPI) (Source: CEPRES). "Before Bubble", "Bubble" and "After Bubble" equal the time frames 1989-1998, 1999/2000 and 2001-2007. p-values are from Mann-Whitney-Wilcoxon-tests. (A Mann-Whitney-Wilcoxon test is a parameter free test which is used to measure the similarity of two independent distributions.) ***, **, * indicate statistical significance at the 1%, 5% and 10% significance level.

Panel A: Gross Internal Rate of Return (IRR) since inception p.a.

	Early		Later		Before Bubble		Bubble		After Bubble		Total	
	EU	USA	EU	USA	EU	USA	EU	USA	EU	USA	EU	USA
Min	-100%	-100%	-100%	-100%	-100%	-100%	-100%	-100%	-100%	-100%	-100%	-100%
25%-Quantile	-100%	-100%	-13%	-75%	-26%	-61%	-100%	-100%	-100%	-100%	-100%	-100%
Mean	-10%	-2%	18%	27%	33%	34%	-26%	-24%	4%	-7%	4%	7%
Median	-41%	-38%	10%	3%	11%	6%	-62%	-72%	-9%	-24%	-7%	-23%
75%-Quantile	13%	18%	35%	45%	41%	51%	3%	-9%	24%	25%	24%	29%
25%-Quantile (weighted)	-100%	-100%	-9%	-59%	-12%	-39%	-100%	-100%	-49%	-87%	-62%	-100%
Mean (weighted)	8%	-18%	18%	19%	65%	30%	-28%	-27%	12%	0%	13%	-1%
Median (weighted)	-26%	-47%	16%	5%	19%	11%	-31%	-60%	0%	-2%	0%	-19%
75%-Quantile (weighted)	11%	11%	39%	40%	55%	43%	6%	-2%	43%	39%	29%	27%
Max	2,250%	2,162%	982%	2,028%	2,250%	2,162%	2,080%	2,028%	1,645%	1,139%	2,250%	2,162%
Standard Deviation	207%	173%	106%	177%	186%	179%	183%	178%	162%	124%	182%	172%
No. of observations	512	2,130	375	1,183	428	1,917	418	1,366	238	763	1,084	4,046
Europe < USA (p-value)	0.06*		0.99		0.93		0.86		0.89		0.89	
After Bubble < Before Bubble (p-value)									0.00***			

Panel B: Total Value to Paid-In (TVPI)

	Early		Later		Before Bubble		Bubble		After Bubble		Total	
	EU	USA	EU	USA	EU	USA	EU	USA	EU	USA	EU	USA
Min	0.00	0.00	0.00	0.00	0.00	0.00	0.00	0.00	0.00	0.00	0.00	0.00
25%-Quantile	0.00	0.00	0.61	0.03	0.33	0.06	0.00	0.00	0.00	0.00	0.00	0.00
Mean	2.22	2.98	2.05	2.84	3.60	4.88	1.01	1.08	1.64	1.32	2.17	2.93
Median	0.20	0.23	1.36	1.09	1.46	1.23	0.10	0.03	0.80	0.50	0.76	0.41
75%-Quantile	1.45	1.82	2.24	2.97	3.00	3.69	1.10	0.71	1.78	1.71	1.93	2.27
25%-Quantile (weighted)	0.01	0.00	0.70	0.07	0.65	0.18	0.00	0.00	0.32	0.02	0.09	0.00
Mean (weighted)	1.62	2.71	2.02	2.56	3.01	5.13	0.95	1.00	1.77	1.58	1.84	2.57
Median (weighted)	0.34	0.18	1.65	1.11	1.97	1.55	0.31	0.06	1.00	0.94	1.00	0.46
75%-Quantile (weighted)	1.45	1.40	3.00	2.82	3.12	3.83	1.23	0.89	2.02	2.30	2.23	2.12
Max	325.53	284.44	25.15	142.89	325.53	284.44	58.32	94.46	45.08	25.82	325.53	284.44
Standard Deviation	14.92	11.77	2.85	7.84	16.35	16.31	3.47	3.90	4.11	2.12	10.73	11.64
No. of observations	512	2,130	375	1,183	428	1,917	418	1,366	238	763	1,084	4,046
Europe < USA (p-value)	0.08*		0.99		0.91		0.92		0.90		0.92	
After Bubble < Before Bubble (p-value)									0.00***			

Table 4.8: Relative performance of fully and partially realized European and US Venture Capital investments into portfolio companies measured in Excess-IRR vs. the MSCI World and Public Market Equivalent (PME) vs. the MSCI World (Source: CEPRES). "Before Bubble", "Bubble" and "After Bubble" equal the time frames 1989-1998, 1999/2000 and 2001-2007. p-values are from Mann-Whitney-Wilcoxon-tests. (A Mann-Whitney-Wilcoxon test is a parameter free test which is used to measure the similarity of two independent distributions.) ***, **, * indicate statistical significance at the 1%, 5% and 10% significance level.

Panel A: Excess-IRR vs. MSCI World

	Early		Later		Before Bubble		Bubble		After Bubble		Total	
	EU	USA	EU	USA	EU	USA	EU	USA	EU	USA	EU	USA
Min	-135%	-142%	-114%	-119%	-117%	-120%	-135%	-112%	-119%	-147%	-135%	-147%
25%-Quantile	-95%	-95%	-17%	-75%	-24%	-60%	-94%	-94%	-98%	-104%	-93%	-94%
Mean	-28%	-12%	15%	15%	16%	21%	-36%	-30%	-8%	-15%	-9%	-3%
Median	-38%	-33%	5%	-2%	4%	0%	-50%	-74%	-16%	-28%	-12%	-23%
75%-Quantile	8%	16%	28%	40%	31%	43%	5%	-7%	23%	21%	18%	25%
25%-Quantile (weighted)	-94%	-95%	-11%	-50%	-14%	-36%	-93%	-94%	-28%	-100%	-49%	-93%
Mean (weighted)	-21%	-23%	15%	9%	17%	20%	-25%	-32%	8%	-9%	-3%	-8%
Median (weighted)	-19%	-35%	11%	-2%	14%	4%	-25%	-52%	-5%	-13%	-2%	-20%
75%-Quantile (weighted)	10%	10%	35%	30%	48%	36%	9%	0%	38%	29%	26%	23%
Max	795%	987%	992%	908%	992%	980%	716%	987%	763%	896%	992%	987%
Standard Deviation	100%	135%	104%	131%	117%	144%	87%	119%	122%	117%	110%	133%
No. of observations	485	2,020	370	1,150	413	1,851	402	1,307	229	728	1,044	3,886
Europe < USA (p-value)	0.12		0.99		0.91		0.73		0.97		0.93	
After Bubble < Before Bubble (p-value)									0.00***			

Panel B: PME vs. MSCI World

	Early		Later		Before Bubble		Bubble		After Bubble		Total	
	EU	USA	EU	USA	EU	USA	EU	USA	EU	USA	EU	USA
Min	0.00	0.00	0.00	0.00	0.00	0.00	0.00	0.00	0.00	0.00	0.00	0.00
25%-Quantile	0.00	0.00	0.48	0.03	0.31	0.05	0.00	0.00	0.00	0.00	0.00	0.00
Mean	1.63	2.29	1.70	2.27	2.52	3.61	1.02	1.14	1.34	1.06	1.68	2.30
Median	0.20	0.20	1.19	0.91	1.12	0.95	0.12	0.04	0.67	0.42	0.61	0.38
75%-Quantile	1.24	1.51	1.93	2.40	2.29	2.79	1.13	0.77	1.42	1.38	1.62	1.91
25%-Quantile (weighted)	0.01	0.00	0.55	0.08	0.39	0.14	0.00	0.00	0.22	0.01	0.08	0.00
Mean (weighted)	1.28	2.01	1.70	2.01	2.18	3.57	1.01	1.06	1.45	1.23	1.51	1.97
Median (weighted)	0.29	0.18	1.36	0.96	1.43	1.12	0.28	0.08	0.88	0.71	0.91	0.44
75%-Quantile (weighted)	1.29	1.28	2.30	2.13	2.48	2.77	1.29	0.93	1.90	1.80	1.93	1.83
Max	174.17	160.52	22.89	86.41	174.17	233.13	48.61	88.01	39.45	18.08	174.17	233.13
Standard Deviation	8.39	8.19	2.39	5.68	9.18	11.91	3.10	3.89	3.50	1.65	6.33	8.62
No. of observations	512	2,130	375	1,183	428	1,917	418	1,366	238	763	1,084	4,046
Europe < USA (p-value)	0.07*		0.99		0.92		0.81		0.91		0.90	
After Bubble < Before Bubble (p-value)									0.00***			

the weighted mean into positive territory (+8%) but the weighted median is still negative at -26%. This discrepancy signals a very skewed distribution of gross IRRs. My results for early stage investments in the USA seems quite similar to Europe in the median but the mean is 8% higher. With a Wilcoxon-test[114] (unweighted) I find that the European mean is statistically significantly lower than the US mean. These results are in line with Gottschalg and Phalippou (2008) and Hege et al. (2009) who find a significant underperformance of European VC to the US counterparts. Beyond that, my data reveal a more detailed picture: Early stage investments underperformed based on the IRR and TVPI measures. However, later stage investments did not underperform; they earned attractive gross IRR levels with a less skewed return distribution both in Europe and in the US. The unweighted mean (median) is 18% (10%) for Europe and 27% (3%) for the USA.

The "Total" data which also include investments declared as "unspecified VC" show that aggregate VC returns in Europe were low between 1989 and 2007. The mean is 4% and the median -7%. The suspicion on fund level by Gottschalg and Phalippou (2008) that European VC is less attractive than US VC is nevertheless not confirmed by my Mann-Whitney-Wilcoxon-tests (unweighted) on portfolio company level.[115] The return difference is not significantly different from zero.

The *CEPRES* sample consists of about 40% investments before the bubble, 40% investments during the bubble (1999/2000) and 20% investments after 2000. I distinguish between these three time periods to uncover differences in the performance over time. My statistics in Table 4.7 show that European and US VC investments which were started before the bubble had

114 A Wilcoxon-test does not require any assumption on the distribution of the variables, otherwise it is very similar to a t-test that compares means of normally distributed variables. As the return figures are very skewed, the Wilcoxon-test is more appropriate than a standard t-test.

115 Although, one could argue that fund level and portfolio company level are not similar the differences should have minor effects on the conclusion. VC funds mostly invest on the same continent (Hobohm, 2008a). Fund management fees, which equal the performance difference between the fund and the portfolio company level, hardly distort the picture as fee structures are almost similar worldwide (Gompers and Lerner, 1999a; Schmidt and Wahrenburg, 2004).

excellent returns. The investments of the years 1999/2000, however, caused massive losses of -26% (-62%) IRR in the mean (median), and the total capital was lost in more than 50% of the investments (see the median TVPI figures in Panel B). After the bubble the returns recovered, but not up to the levels before the bubble. A Mann-Whitney-Wilcoxon-test for a negative difference between the IRR distribution before and after the bubble is statistically highly significant (p-value: 0.00***). IRRs after 2000 were at -9% at the median and +4% at the mean. This positive mean is the consequence of some exceptional investment returns. A Mann-Whitney-Wilcoxon-test indicates no statistically significant difference between Europe and the US in any of the three time periods. For the time before the bubble this is surprising because Figure 4.2 based on VentureXpert data suggested a significant difference in the performance between Europe and the US on fund level for the vintage years 1994-1998. I have no clear explanation why the European fund performance in VentureXpert for 1994-1998 is so low in comparison to the US fund performance while the portfolio company performance hardly differs in the *CEPRES* data.

The TVPI statistics in Table 4.7 Panel B confirms my findings on the gross IRRs. The high skewness in returns becomes even more obvious: the median TVPI is low at just 0.2 while the mean at 2.22 seems attractive.

Tables 4.8 Panel A and B display a comparison to the benchmark index MSCI World. I chose the MSCI World as it is a global index which allows me to compare VC performance on both sides of the Atlantic without distortions from regional specifics. Early stage investments in Europe and the USA underperformed the benchmark MSCI World while later stage investments outperformed between 1989 and 2009 overall. However, there is a significant difference in the time spans before and after the bubble. While VC investments before 1998 showed an excellent average outperformance of 16% in comparison to the MSCI World, this asset class underperformed by -8% after 2000. Investments which were started in the bubble underperformed completely; the low standard deviation during this period (1999/2000) confirms the massive losses across the board. US investments show similarly bad figures.

The PME statistics are not as negative as the Excess-IRR numbers. Although the median PME points to a significant underperformance compared to public markets at 0.67 after the bubble, which is significantly worse than before the bubble, fund managers obviously found a few totally exceptional portfolio companies which pulled the mean PME into positive territory. This spread between mean and median is even more obvious in the "Total" statistics.

IRR Development by Investment Year Figure 4.5 displays the development of gross IRRs of the European VC investments from the *CEPRES* sample by investment years. Median returns (the fat lines in the box plots) reveal positive returns between 1992 and 1998. Thereafter, median returns dropped significantly to almost -100% in 2000, recovered into positive territory for the investment years 2003/2004 and fell sharply in the last few years again. Thus, the two years 2003 and 2004 were the only attractive investment years in the aftermath of the Internet bubble. This is not surprising, as Section 4.6 will illustrate. Fund managers should have started a main share of their investments in these years (2003 and 2004) to realize attractive fund returns (which did not happen in reality).

Returns in Different Countries and Industries The economies of the European countries developed quite differently in the recent years. To see if investment returns are heterogeneous in the main countries in Europe I compare returns of investments by VC funds, i.e., United Kingdom, France and Germany, with a summary including the rest of the countries, in Figure 4.6. Germany steps out in this chart; some exceptional investment returns pull the mean and weighted mean far into positive territory in the overall time period. The outperformance of Germany in comparison to the other European countries is consistent over time: The unweighted mean before, in

Figure 4.5: Gross IRRs since inception of European VC investments by investment years (equals the year of the first investment) (Source: CEPRES)

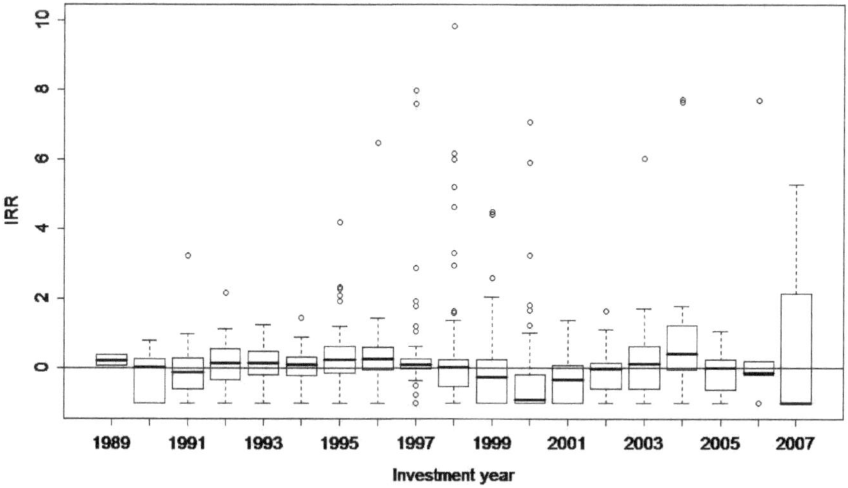

and after the bubble is the highest in the reference group. UK and France performed far worse and negative especially in the bubble and its' aftermath.

A comparison of industry sectors (Figure 4.7) reveals that Telecommunication and Computer were the best performing sectors before the bubble but suffered heavily in the bubble. I associate these sectors' before-bubble-performances with the excellent exit market opportunities in 1999/2000 coinciding with exceptionally high valuations in these sectors.

Another interesting finding is about investments in Industrials: this sector performed consistently well before the bubble, least suffered in the bubble and shows a good performance afterwards. Good median returns of more than 10% were even realized after 2000. All other sectors – except from the Telecommunication sector which had very few extraordinary investment returns after 2000 – performed negatively in the bubble and the aftermath. I conclude that conservative sectors seem to allow more robust positive returns over time while "hype" sectors live from market irregularities like the Internet bubble.

Figure 4.6: Comparison of Gross IRRs since inception of fund investments into portfolio companies from different European countries (Source: CEPRES)

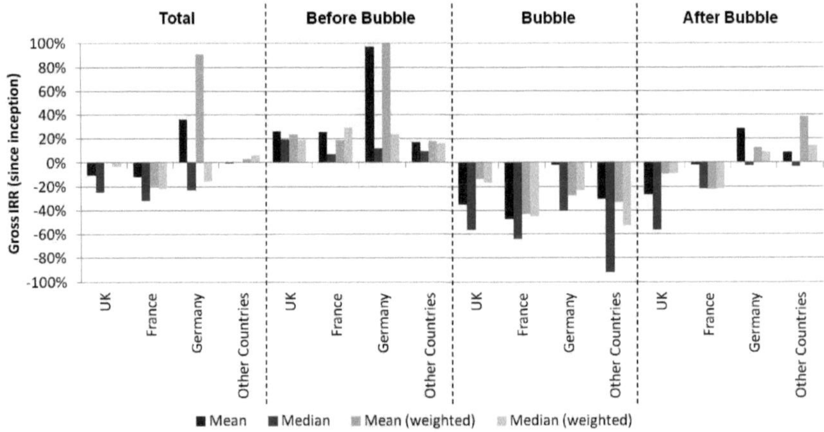

Figure 4.7: Comparison of Gross IRRs since inception of fund investments into portfolio companies belonging to different industry sectors (Source: CEPRES)

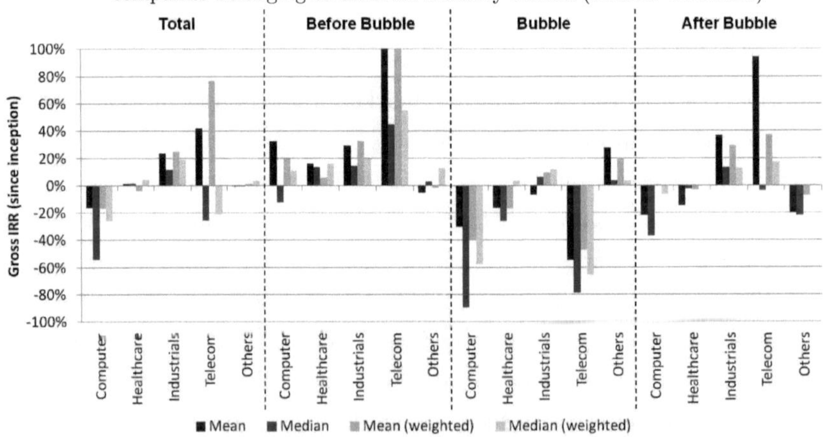

4.5 Investor Success in European Venture Capital

4.5.1 Overview on Limited Partner Sample

The previous section has shown that the average performance of European Venture Capital was weak after 1998. This does not necessarily exclude that some groups of LPs with superior fund selection skills and the "right" investment strategy chose the "right" investment time and the "winner" funds and made a significantly better return than the market average. Other authors have found that fund selection is central to a good performance in VC investing due to the strong diversity in fund returns (e.g., Kaplan and Schoar, 2005).

Before going into detail on the investment behavior and respective success of different investor groups I investigate the composition of the investor structure in the fund sample descriptively. The investors who provided the data on European VC funds also gave me the names of the LPs who invested in the same VC funds. Nevertheless, I received only a partial allocation of the LPs to investor groups. To complement the information I conducted detailed Internet research and were able to validly identify the investor type of 88% of the 1,262 investors in the sample. The unidentifiable rest of the investors is allocated to the group "Other/Unidentified".[116]

To determine the representativeness of my LP structure I compare it to the LP structure of a survey on the European VC industry from the EVCA, Thomson Financial and PWC. Table 4.9 shows my results. I find that government agencies hold a significantly higher share in my sample than in the market average. This may cause distortions; however, I cannot control them. Though, I expect the relative considerations, which I conduct in this section, to be affected only minor. The high share of government agency investments implies that the shares of the other LP groups are lower

116 I also include the group "Advisors" in the group "Other/Unidentified" as I found too few records for the advisors to allow clear statements.

Table 4.9: Comparison of the shares of different LP groups in the funds of the vintage years 1999-2002 between the restricted fund sample and results from an EVCA/Thomson Financial/PWC survey

	Fund sample				EVCA/Thomson Financial/ PWC survey			
	1999	2000	2001	2002	1999	2000	2001	2002
Banks/Asset manager	22%	15%	18%	16%	31%	22%	24%	26%
Corporations	12%	8%	8%	9%	10%	11%	6%	7%
Endowments/Family Offices/ Trusts/Foundations/Individuals	4%	7%	10%	3%	7%	8%	9%	8%
Funds-of-funds	3%	5%	3%	8%	4%	11%	12%	13%
Government agencies	20%	15%	20%	32%	5%	6%	6%	11%
Insurances	11%	10%	6%	10%	14%	13%	12%	14%
Other/Unidentified	14%	38%	25%	22%	8%	5%	4%	4%
Pension funds	13%	3%	9%	0%	20%	24%	27%	16%

accordingly. Moreover, the high share of LPs in the "Other/Unidentified" group in my sample (especially in the vintage year 2000) requires a careful comparison of the distributions in the two samples.

It seems that especially the shares of the three types of financial institutions Banks/Asset managers, Fund-of-funds as well as Pension funds is lower in my sample than in the overall market over time. The results of the next analysis of LP behavior and success should be interpreted behind this background. I assume to have one of the best samples concerning this question currently available in the European market, but I cannot truely rule out selection biases in the LP data.

For the rest of this investigation I consider each LP-fund-combination as an observation and control for the fact that multiple LPs respectively invest in one fund by considering the clustering in the standard errors of my regressions later on. Table 4.10 gives a descriptive overview on the behavior and success of LPs in European VC funds. Panel A shows the number of LP-fund-relationships in my full fund sample up to the year 2005. I use the full sample for this investigation as I am just interested in the relative performance between LP groups and can control for distortions due to the different ages of the funds by specifying year-fixed effects in my regression analysis. All investor groups in my sample have 75 and more investments into European VC funds. This amount should allow for serious conclusions from the data. The groups with a minor number of investments in my sample were either grouped like Endowments, Family Offices, Trusts, Foundations and Universities or allocated to the "Other/Unidentified" group. The group around Endowments was combined because anecdotal evidence suggests that these investor types have a rather similar size and investment interests. Their main distinction from other investor classes like Banks, Insurance companies or Funds-of-funds is that they all invest their own capital and not capital belonging to others.

The most LP-fund-combinations in my sample are from Government agencies who invested 23.3% of the total capital, Banks and asset managers with 16.9% of committed capital and the "Other/Unidentified" group with 18.5% of capital. The least capital was invested by GPs of funds (1.9% of committed capital), the group around Endowments (2.7%) and from Individuals (1.6%).

Turning to the shares of commitments which go to different stages, the overall split is 70% for early stage, 19% for later stage and 12% for balanced stage. Because early stage commitments usually have smaller sizes the split by capital is 30/38/32 for early, later and balanced stage.

The group around Endowments invested preferentially into early stage VC (41%). Individuals (40%) and Pension Funds (39%) follow directly. Corporations (consisting mainly of chemical, pharmaceutical as well as technology companies) invested the least (just 21%) into early stage VC, but committed to relatively many later stage funds. This points to the fact, that Corporations have more than just capital return intensions with an investment into VC: They want to get insight into many new businesses without risking much capital. Attractive portfolio companies are potential acquisition targets for the Corporations.

Later stage funds were least financed by Individuals (11%) and the group around Endowments (22%). GPs of funds seemed to commit preferably to later stage funds and less to high risky early stage funds. A potential reason seems the diversification rationale: Fund managers whose return on their human capital strongly depends on the performance of their funds anyway are very reluctant to additionally invest their personal assets into these funds. They obviously prefer more predictable fund structures like later stage funds over early stage funds.

Panel B shows the investment success of the LP groups along four performance measurements. The relative performance is quite stable along the different measures. Corporations had the worst performance with -12% mean IRR and a mean TVPI of 0.65; the group around Endowments was almost similarly bad. This is surprising as I find exactly the contrary to the results from Lerner et al. (2005) who investigated the time-frame 1991-1998 for the USA. Though, my results confirm the findings of Hobohm (2008b) for Endowments for the same time-frame as ours.

In my data sample, the "Other/Unidentified" had a surprisingly strong performance together with GPs of funds and Government agencies (all -7% mean IRR). The other LP groups lie in the middle.

4.5.2 The Investment Behavior of Different Investor Groups

In the descriptive analysis I am not able to exclude environmental changes over time - especially possible distortions from different fund ages[117] - and other factors that potentially blur the real picture. So I turn to regression analysis for a more detailed investigation of LP behavior and success. To investigate the preferences of the different LP groups for the more or less risky VC stages I use multinomial logit regression similar to an analysis by Hobohm (2008a). This regression type estimates the odds of an investment being in one of the stage categories conditioned on independent variables. I include year-fixed effects to control for changes in the market environment over time that I cannot approximate with independent variables. My reference stage category in the two specifications of Table 4.11 is "early stage".

117 Treating young and mature funds equally in the analysis could lead to wrong conclusions. While return figures of mature funds are close to their final returns, young funds usually have returns below their final returns as they didn't exit many of their investments yet.

Table 4.10: Descriptive overview on LP behavior and success in the full fund sample
The table shows descriptive statistics on 1,781 investments from the full fund sample grouped by LP classes. The fund sample contains data on funds raised between 1994 and 2005 with 96% of the funds from 1998 and later. Performance statistics in Panel B are net of carry and management fees.

Panel A: Investment behavior of LPs in European VC funds

	No. of investments	Share of committed capital	Average commitment (€M)	**Share of no. of commitments to funds with stage focus**			**Share of committed capital to funds with stage focus**		
				Early stage	Later stage	Balanced stage	Early stage	Later stage	Balanced stage
Banks/Asset manager	292	16.9%	5.6	77%	14%	10%	25%	37%	38%
Corporations	144	6.2%	4.1	89%	5%	5%	21%	41%	38%
Endowments/Family Offices/Trusts/ Foundations/ Universities	81	2.7%	3.2	68%	18%	14%	41%	37%	22%
Fund-of-funds	107	7.9%	7.1	76%	13%	11%	36%	38%	26%
Government agencies	281	23.3%	8.0	67%	20%	13%	28%	43%	29%
GPs of Fund	75	1.9%	2.4	75%	16%	9%	33%	21%	46%
Individuals	189	1.6%	0.8	81%	14%	5%	40%	49%	11%
Insurances	154	7.1%	4.5	68%	17%	15%	32%	41%	27%
Other PEFs	154	7.3%	4.6	77%	9%	14%	29%	44%	27%
Other/Unidentified	196	18.5%	9.4	64%	24%	13%	35%	30%	34%
Pension Funds	108	6.5%	5.8	63%	17%	20%	39%	29%	32%
Total/Mean	1,781	100.0%		70%	19%	12%	30%	38%	32%

Panel B: Investment success of LPs in European VC funds

	IRR		TVPI		Excess-IRR vs. MSCI World		PME vc. MSCI World	
	Mean	Median	Mean	Median	Mean	Median	Mean	Median
Banks/Asset manager	-9%	-10%	0.76	0.71	-6%	-5%	0.84	0.83
Corporations	-12%	-12%	0.65	0.59	-8%	-5%	0.75	0.70
Endowments/Family Offices/Trusts/ Foundations/ Universities	-12%	-11%	0.66	0.61	-8%	-9%	0.80	0.76
Fund-of-funds	-8%	-9%	0.80	0.73	-1%	-1%	0.99	1.07
Government agencies	-7%	-10%	0.80	0.73	-2%	-5%	0.92	0.84
GPs of Fund	-7%	-8%	0.82	0.76	-2%	-4%	0.94	0.95
Individuals	-9%	-9%	0.74	0.68	-6%	-5%	0.82	0.78
Insurances	-8%	-8%	0.79	0.73	-3%	-2%	0.91	0.90
Other PEFs	-10%	-10%	0.71	0.68	-4%	-5%	0.86	0.80
Other/Unidentified	-7%	-8%	0.84	0.83	-1%	-3%	0.98	0.98
Pension Funds	-11%	-12%	0.68	0.64	-7%	-8%	0.80	0.75

Thus, the odds have to be interpreted as the increased or decreased chance to select a certain fund stage over an early stage focussing fund.

Table 4.11: Investment behavior of different groups of limited partners (LPs)
The presented analysis is based on the full fund sample. I calculate two specifications of Multinomial Logit regression models which provide the likelihood of picking a certain fund type – in comparison to early stage focussed funds – depending on independent variables. Specification 1 includes LP dummies and Specification 2 tests for the effects of different fund sizes and fund manager experience (measured with the consecutive fund number in a series of all funds of the same fund manager). All estimations include vintage year fixed effects. All observations were clustered at the fund level, robust standard errors are reported in brackets. ***, **, * indicate statistical significance at the 1%, 5% and 10% significance level. A constant effect was estimated but is not reported. Base categories for comparison: Early stage funds for the dependent variable; Corporations for the LP dummies.

Multinomial Regression (Dependent Variable: Fund Focus Stage (Early Stage is reference category))

	Specification 1		Specification 2	
	Balanced stage	Later stage	Balanced stage	Later stage
Banks/Asset manager	0.83*	0.34	0.82*	0.29
	(0.44)	(0.41)	(0.45)	(0.39)
Endowments/Family Offices/Trusts/ Foundations/Universities	1.05**	0.68	0.89*	0.42
	(0.48)	(0.61)	(0.48)	(0.45)
Funds-of-funds	0.61	0.65	0.55	0.59
	(0.41)	(0.63)	(0.41)	(0.67)
GPs of Fund	0.54	0.45	0.49	0.32
	(0.42)	(0.40)	(0.43)	(0.37)
Government agencies	0.92***	0.64**	0.87***	0.51*
	(0.31)	(0.30)	(0.32)	(0.29)
Individuals	0.92	-0.63	0.97	-0.57
	(0.75)	(0.69)	(0.75)	(0.68)
Insurances	1.38***	0.94**	1.29***	0.73
	(0.42)	(0.46)	(0.44)	(0.48)
Other PEFs	0.95***	0.50	0.93***	0.50
	(0.36)	(0.43)	(0.36)	(0.43)
Other/Unidentified	0.56	-0.21	0.53	-0.27
	(0.39)	(0.41)	(0.39)	(0.40)
Pension Funds	1.28**	1.59***	1.11**	1.18***
	(0.51)	(0.45)	(0.54)	(0.37)
Fund size (log)			0.32	0.04
			(0.36)	(0.53)
Fund mgr. Exp.			0.06	0.09*
			(0.05)	(0.05)
Vintage Year Fixed Effects	Yes		Yes	
No. of cluster	137		137	
No. of observations	1,781		1,781	
Pseudo R^2	18.1%		20.3%	

In Specification 1 I first investigate the stage preferences of the different LP types in comparison to the reference LP type "Corporations". All investor types seem to have a higher preference for balanced stage funds than Corporations. However, the preference is just statistically significant for Banks/Asset managers, the group around Endowments, Government agencies, Insurances, other PE funds and Pension funds. The fact that Corporations have a low preference for balanced funds confirms my hypothesis that these investors have other agendas than just generating investment returns. For Fund-of-funds the relatively low preference for balanced funds is equally not surprising as these funds are mixing asset classes themselves as well and do not require another fund manager to mix them. The only reason I see why a Fund-of-funds might want to invest in a balanced stage fund is to get access to the fund's deal flow which can be extraordinary compared to other funds.

In relative terms Corporations seem to have a low preference for later stage funds; just Individual VC investors and the "Other/Unidentified" group might prefer these funds less (but not statistically significant). Government agencies, Insurance companies and Pension funds had a statistically significant higher attachment to later stage funds than Corporations.

An interesting finding is that Individuals mostly prefer investing into early stage funds when engaging in Venture Capital. This shows that these people have a high preference for risk. Moreover, the Individuals may not be able to invest in later stage funds: later stage funds require and usually have a higher volume. Thus, Individuals who just invest €800K on average are probably not a later stage funds' most preferred investors unless the managers are desperately searching for capital. The relatively low preference for later stage funds of the group around Endowments is in line with my expectations. Especially Endowments have the image to be highly entrepreneurial.

Specification 2 is similar to Specification 1, but additionally includes the fund size and the fund managers' experience in the regression. The consecutive number of a fund in the series of all funds of the same fund manager is used as an indicator for fund manager experience. I find that my results obtained in Specification 1 are robust and hardly influenced by size or experience concerns. Just the preferences from Insurance companies for later stage funds turns insignificant. The fund manager experience does not seem to affect the decision if a balanced stage fund is chosen, but the later stage fund selection is positively affected. VC fund managers often evolve from early stage investing to later stage investing over time because more experience usually goes along with bigger fund sizes. Thus, fund managers need to search for investment opportunities with higher capital requirements which they find in later stage investments. So, it is easier for fund managers to get capital from many investors for their later stage funds if they have more experience.

To further test the robustness of my results I also specified the analysis excluding fixed effects. The results are very similar except for the statistical significance for the LPs' preferences for balanced stage funds: The Banks/asset manager coefficient looses its' significance but the coefficients for Fund-of-funds, GP of funds and "Other/Unidentified" become significant. For Fund-of-funds this is surprising but could be due to differences in the preferences in the market cycles.

Overall, I find only quite minor differences between the investors' preferences within the asset class Venture Capital. It seems the decision in VC investing is less about whether to choose early or later stage, but more about whether to invest in VC at all (in comparison to investing in buyout or even other asset classes).

4.5.3 The Investment Success of Different Investor Groups

With the quite minor differences in the investment behavior in mind, I expect only minor differences in the investment performance of the different LP types. This is confirmed by my regression analysis in Table 4.12. The table shows an OLS regression analysis on LP-fund-combinations as observations. The IRR of an LP's respective fund investment - i.e., the fund IRR since inception - is the dependent variable. Explanatory variables are the LP type dummies, stage dummies, the size of the fund which the LP invested in, the fund manager's experience and interaction variables between LP type and stage dummies. All specifications include vintage year fixed effects to remove possible sources of bias from different fund maturities in my full fund sample. Funds with different vintage years naturally have different chances and risks; this influence needs to be excluded.

The maximum performance difference between the different LP types is 4.45% IRR in Specification 1. This result is low in comparison to the results by Lerner et al. (2005) and Hobohm (2008b) for the time-frame 1991-1998. The authors find differences of up to 16% between the worst and best investor group. However, I saw in Figure 4.2 that the overall performance in the market decreased heavily after 1998 worldwide. This is confirmed by Hobohm (2008b) who states that the differences he finds for 1991-1998 decrease to 7% for 1999-2003. Thus, I conclude that investors may distinguish from each other in good times but equally suffer in bad investment times. Fund managers after 1998 seemed to behave quite similarly and obtained bad returns on their investments.

Corporate investors were performing worst in my full fund sample. Possibly these investors do not even mind the weak fund performance as they probably want to get insight into the latest technology or biotechnology research of the start-ups and do not focus their attention on the direct investment return. Banks/asset managers, GPs of funds, Government agencies, Individuals and "Other investors" statistically significantly distinguished from Corporate investors. The result on Government agencies in my sample does not confirm the anecdotal evidence that Government agencies allegedly were terrible investors – at least not after 1998.

It seems later stage funds might be doing a little better than early stage funds (the reference group) and those again did better than balanced stage funds. However, the performance differences between VC funds which focus on different stages is not statistically significant.

Table 4.12: Investment success of different groups of limited partners
This table reports pooled OLS regressions of net IRRs per fund on dummy variables for LP classes, fund stage classes and covariates. All specifications include vintage year fixed effects. Specification 1 is a simple regression using LP dummies. Specification 2 includes LP dummies, fund size (logarithmized) and fund manager experience (measured with the consecutive fund number in a series of all funds of the same fund manager). Specification 3 shows interaction terms for LP classes with fund stage classes. Robust standard errors are reported in brackets; I clustered all observations at the fund level. ***, **, * indicate statistical significance at the 1%, 5% and 10% significance level. Base categories for comparison: Early stage funds for the fund stage classes; Corporations for the LP classes.

Linear Regression (Dependent variable: IRR (in %))			
	Specification 1	**Specification 2**	**Specification 3**
Banks/Asset manager	3.67**	3.71**	5.03***
	(1.49)	(1.51)	(1.67)
Endowments/Family Offices/Trusts/	2.31	3.09*	2.78
Foundations/Universities	(1.97)	(1.84)	(2.00)
Fund-of-funds	2.89	3.22	4.59*
	(2.00)	(2.06)	(2.44)
GPs of Fund	3.96**	4.14**	3.90*
	(1.99)	(1.97)	(2.36)
Government agencies	4.26**	4.45**	4.70*
	(2.14)	(2.13)	(2.70)
Individuals	4.41**	4.15**	4.06*
	(2.00)	(2.01)	(2.21)
Insurances	3.41	3.84*	5.65**
	(2.16)	(2.17)	(2.69)
Other PEFs	2.38	2.44	2.14
	(1.57)	(1.61)	(1.93)
Other/Unidentified	4.45**	4.60**	4.47**
	(1.79)	(1.83)	(2.12)
Pension Funds	1.22	2.03	2.43
	(1.94)	(1.93)	(2.76)
Later stage	1.14	2.01	5.12
	(3.61)	(3.60)	(3.73)
Balanced VC	-2.87	-2.12	-0.73
	(2.74)	(2.73)	(4.26)
Fund size (log)		-1.67	-1.82
		(1.21)	(1.15)
Fund manager experience		-0.28	-0.28
		(0.21)	(0.20)
Banks/Asset manager * Later stage			-10.02***
			(3.34)
Banks/Asset manager * Balanced VC			-3.35
			(3.49)
Endowments/Family Offices/Trusts/			-7.84
Foundations/Universities * Later stage			(4.80)
Endowments/Family Offices/Trusts/			5.45
Foundations/Universities * Balanced VC			(5.31)
Funds-of-funds * Later stage			-8.94*
			(4.54)
Funds-of-funds * Balanced VC			-1.89
			(5.37)
GPs of Funds * Later stage			-0.55
			(3.94)
GPs of Funds * Balanced VC			1.64

Continued on Next Page...

	Specification 1	Specification 2	Specification 3
			(4.65)
Government agencies * Later stage			-0.67
			(4.38)
Government agencies * Balanced VC			-2.00
			(3.97)
Individuals * Later stage			8.76
			(5.51)
Individuals * Balanced VC			-1.92
			(4.39)
Insurances * Later stage			-2.58
			(4.13)
Insurances * Balanced VC			-7.55*
			(3.86)
Other PEFs * Later stage			-0.36
			(3.60)
Other PEFs * Balanced VC			1.20
			(3.06)
Other/Unidentified * Later stage			0.98
			(5.44)
Other/Unidentified * Balanced VC			0.50
			(4.72)
Pension Funds * Later stage			-6.01
			(4.18)
Pension Funds * Balanced VC			2.28
			(5.77)
Constant	-11.63***	18.96	21.13
	(2.08)	(21.28)	(20.35)
Vintage Year Fixed effects	Yes	Yes	Yes
No. of cluster	135	135	135
No. of observations	1,778	1,778	1,778
R^2	14.8%	16.8%	18.4%

Bigger funds pointed to a worse performance than smaller ones (see Specification 2), however, this result is not statistically significant. When controlling for this fact, the group around Endowments and Insurance companies also invested significantly better than Corporations. Interestingly, funds of more experienced fund managers seemed to underperform younger fund management teams – yet, statistically not significant. This result is contrary to the findings of Gottschalg and Phalippou (2008) who confirmed the intuitive assumption that funds of experienced managers perform better. The regression specification 3 with interactive effects between LP types and stages reveals that Banks/asset managers were investing worse than corporate investors in later stage funds, but compensated this low performance with a superior investment performance in early stage funds. The good performance of Funds-of-funds originates from their early stage investments; they invested badly in later stage funds. The overall good performance of Insurance companies was negatively affected by their investments into balanced stage funds.

To check the robustness of my regressions I conducted multiple quality checks: In a first set of regressions I left out the Top-1% and the Bottom-1% of the funds. This implied that the performance of GPs of funds and Government institutions were not statistically significant any more, but Banks/Asset manager, Individuals and the "Other/Unidentified" groups remained statistically significant and positively distinguished from Corporate investors. In another set of regressions I tested Excess-IRR in comparison to the MSCI Europe as the dependent variable. The results were economically unchanged except for Banks/Asset manager. The regression coefficients for Banks/Asset manager loose statistical significance.

A direct comparison of my results to existing studies, e.g., Lerner et al. (2005) and Hobohm (2008b), is hampered as their investigations do not allow final conclusions on the time period after 1998 for European VC. Yet, some regression specifications by Hobohm (2008b) enable directive insight into the LPs' investment success in European VC: His cross-country analysis reveals that generally early stage VC funds had a bad performance worldwide and later stage VC funds did so especially in Europe. In his specifications checking the differences between the US, Europe and the rest of the world, he does not uncover major differences between the regions. Insurances, Endowments and Family Offices seem to have performed better than other investor classes worldwide and Funds-of-funds underperformed especially in Europe.

My results are similar to Hobohm's findings for the USA that the success difference between LPs after 1998 is much lower than before the bubble. This points to the fact that investing in the bubble period and the aftermath was quite different to the time before. Historic investment concepts had to be adapted. Old winners partially became losers and vice versa.

4.6 Reasons for the Low Performance of European Venture Capital

In the previous sections I have found that European Venture Capital was not an attractive asset class to invest in during the last decade - for none of the 11 LP groups which I investigate. This is a little surprising on the first sight in comparison to the two decades prior to the bubble. Several authors found that the worldwide VC market had an attractive performance during the 1990's; as mentioned in Section 4.2, some authors even find an outperformance of VC in comparison to public markets. In this section I uncover potential reasons for the industry's negative track record after 1998. Therefore, I look at the drivers of a VC market's development.

Gilson (2003) states that a well functioning VC market needs the simultaneous availability of entrepreneurs, investors with a taste for risk, and specialized financial intermediaries. He declares that these factors are interdependent and if just one component is missing it emerges automatically. The market environment and conditions also play an essential role for the development of a VC market (Bonini and Alkan Aktuccar, 2009; Romain and van Pottelsberghe de la Potterie, 2004; Jeng and Wells, 2000; Gompers and Lerner, 1999b) - especially the stock and IPO markets (Black and Gilson, 1998). The market environment heavily influences the success possibilities of fund managers. This affects their chances to raise funds from investors in a second step. Moreover, entrepreneurs more prone to start their own business, if they expect a chance to sell it successfully afterwards. Thus, I investigate four topics that potentially contributed to the low performance of European VC in the time from 1999 to today[118]:

1. attractiveness of the market environment in the time-frame (GDP growth, stock market development, IPO market development, etc.),
2. availability of innovative investment projects,

118 According to Gilson (2003) contractual relationships are also important for the cooperation of investors, fund managers and portfolio companies and thus for the evolution of the market. However, contractual terms are fairly uniform internationally (Schmidt and Wahrenburg, 2004).

3. availability of investors with sufficient amounts of capital and skills to select good fund managers,
4. availability of a sufficient level of skills by fund managers to invest successfully.

4.6.1 European Market Attractiveness

Venture Capital funds invest in young companies during their seed, start-up and development phases. They hold the portfolio companies just for a limited period of time (usually a few years) before they sell them or write them off (Sahlman, 1990). In this business model, the fund managers require an environment of low valuations for investing, to be able to buy stakes in portfolio companies at an attractive price (Gompers and Lerner, 2004). The best conditions to successfully develop the young companies is obviously provided by a growing economy in the portfolio company's target market during the holding time of an investment. For a successful exit, fund managers need capital market conditions that allow trade sales or IPOs - if possible at high valuations.

Black and Gilson (1998) identify a well functioning IPO market as a key for the success of Venture Capital. Portfolio companies sold through IPOs are generally associated with higher returns (Schwienbacher, 2005). As the main European countries built technology-oriented stock markets in the late 1990's, a missing stock market may hardly serve as an argument for the bad VC performance (see also Hege et al., 2009), however liquidity in the IPO market might have been a problem.

Stock and IPO market conditions I measure the attractiveness of entry and exit conditions into European portfolio companies by looking at the 12-month-change of the MSCI Europe. Figure 4.8 shows the development between 1990 and 2009. After the burst of the Internet bubble in 2000, capital markets retreated significantly between 2001 and 2003. After the exaggeration in company valuations in the bubble, valuations returned to normal or even understated levels in 2002 and 2003. Between 2002 and 2004

Figure 4.8: Development of 12 month-change of MSCI Europe between 1990 and 2009 (Source: Datastream)
The chart shows monthly observations for how the MSCI Europe developed in the respective previous 12-month-periods.

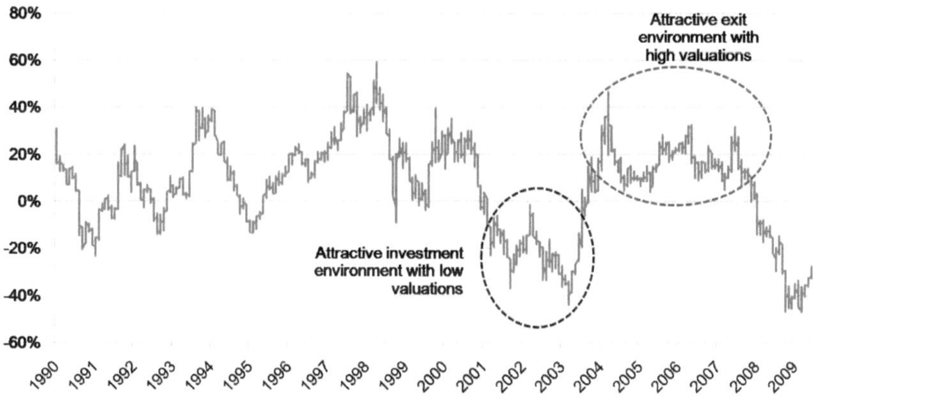

valuations seemed to have been sufficiently low for fund managers to invest at attractive conditions.

History shows that stock markets quickly recovered in the years after 2003. In 2007 the MSCI Europe reached the same levels as in the peak of the Internet bubble. This shows that valuations recovered to attractive levels to sell portfolio companies, e.g., in 2006/2007. The IPO market confirms this view in the years 2004-2007 (see Figure 4.9). Market conditions were good enough to list new companies at the stock market. However, it is interesting to note that the PE funds organized about 35 IPOs per quarter in these years which lies at the lower end of the number of IPOs between 1991 and 2000. This is surprising. Why didn't the fund managers sell more companies between 2004 and 2007? I try to answer this question later in the section about the investment behavior of VC fund managers (Section 4.6.4).

I ascertain that the time after 1998 (bubble and aftermath) had one attractive market cycle for fund managers to earn attractive returns. This

Figure 4.9: Quarterly data for the number of Private Equity (Venture & Buyout) backed IPOs in Europe between 1990 and 2009 (Source: VentureXpert)

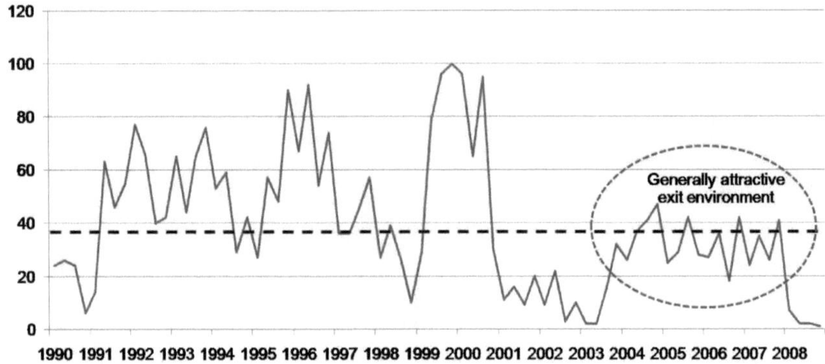

growth cycle started in 2003 and ended in 2008. It lasted about five years which seems sufficient for a portfolio company to start and develop its' business to a point where a fund manager can sell it. Ljungqvist and Richardson (2003) find an average (median) holding period of 3.6 (3) years for PE investments of funds with vintage years 1981-2001. Cumming and MacIntosh (2001) find that the average duration of PE funds' investments started between 1992 and 1995 in the US was 4.75 years. It took on average 4.7 years from investment to IPO and 5.1 years from investment to an M&A deal.

In Figure 4.5 (page 187) I investigated the gross performance of VC investments over time. The box-plots for 2003/2004 show that these years offered good return opportunities. Thus, capital market conditions were in favor for fund managers.

Economic growth in Europe There is a question, if the growth opportunities from 2003 to 2008 have also been in favor for the GPs, i.e., supportive for their portfolio companies? To answer this question I consider the GDP growth in European economies over that time. GDP growth has been identified as a parameter of fund performance by several studies (Kaserer and Diller, 2007; Romain and van Pottelsberghe de la Potterie, 2004; Gompers and Lerner, 1999b). It is expected to positively affect fund returns, i.e., the

Figure 4.10: Quarterly GDP growth (over 12 month periods respectively) in the six European Countries with the highest representation in the CEPRES sample (Source: Eurostat)

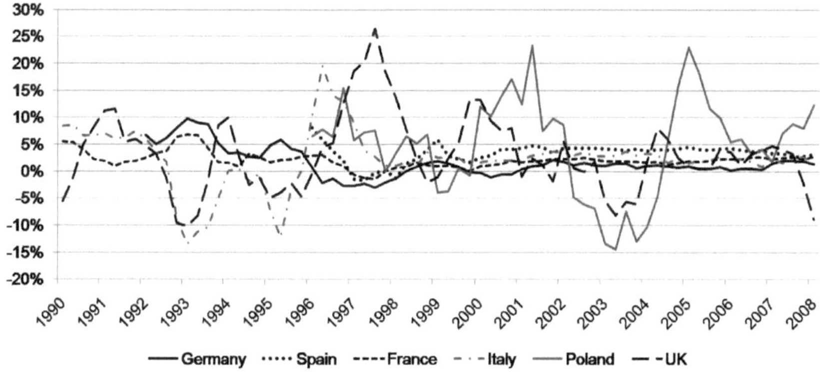

Figure 4.11: Business Climate Index (BCI) in Europe between 1990 and 2008 (Source: Thomson One Banker)

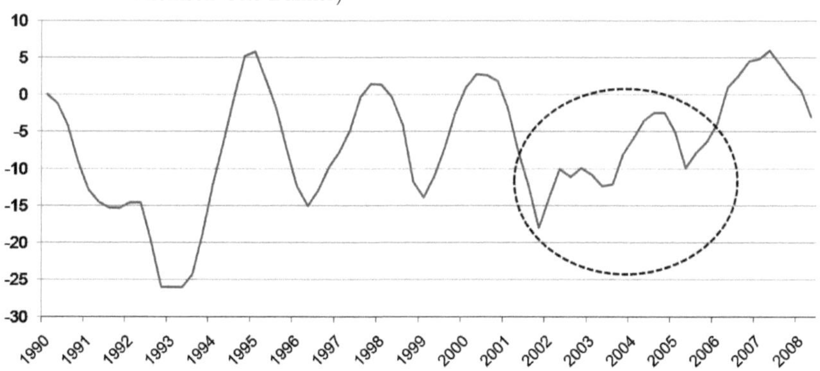

chances for success rise when the economy is growing while a fund holds its investments. Figure 4.10 shows the GDP development between 1990 and 2008 for the six European economies that are most represented in the *CEPRES* sample.[119] Interestingly, the volatility of the economic growth is higher during the 1990's than after the year 2000 for the countries Germany, France, Spain, Italy and UK. Just Poland experienced times of extremely negative and positive growth in excess of +/-10% in the bubble aftermath. Germany, France, Spain and Italy had quite constant, but low growth rates of 0-5% between 2001 and 2008. This indicates that the GDP development was relatively predictable, however not grandiose for young companies. Thus, one reason why VC investments did badly in that time frame may be attributed to the slow growth in the portfolio companies markets which also provided only insufficient chances for them to grow.

Figure 4.11 shows the historic development of the Business Climate Index (BCI). The BCI is composed of five components: a) production trends in recent months, b) order books, c) export order books, d) stocks and e) production expectations. It measures the trust of market participants in the market development. The index is considered a forward looking indicator for economic growth.

My statistical analysis confirms the lack of trust that market participants had in the economic recovery in 2003/04. Despite the consistently positive development in 2004 and later, the index did not reach positive territory until 2007. It seems that the burst of the Internet bubble heavily destroyed trust in the economy. It took up to 7 years until people and investment professionals started to believe in a European "growth story" again. Unluckily, their trust returned right before the market collapsed due to the financial crisis in 2008.

119 The representation is considered as a measure for the importance of an economy for the European VC market.

4.6.2 European Innovativeness

Good business ideas are the most important factor for Venture Capital success. Only if young portfolio companies are successful, investors can earn attractive rates on their capital as well. As Meyer (2006) states, "the correlation between VC and measures of entrepreneurial activity is positive and statistically robust". Thus, another reason for the negative performance of European VC could have been the lack of attractive investment projects in the time after 2000.[120]

I can measure innovativeness only indirectly: The number and quality of start-up companies which are searching for capital is generally unknown. I need proxies for entrepreneurship and innovation to give a qualitative answer[121]: For entrepreneurship I use the founding rate. It equals the ratio of businesses founded in a year to the number of existing companies. Of course, this measure is just indicative, because all new founded businesses are recoded in the statistic – from the bakery next door to a joint venture of big companies. I am aware that this founding rate cannot stand proxy for the number of potentially attractive start-ups for VC funds; however, this measure should allow me to separate times when few start-ups are founded from times with increased activity.

Figure 4.12 shows the founding rate for four main European countries[122] between 1998 and 2006. It seems that UK is more entrepreneurial than Spain, France and Italy, with Italy having the least company foundations. Especially interesting for my analysis is the development over time. The share of newly established companies slightly decreased in the years 2001 and 2002 - right after the bubble. However, the levels quickly picked up

120 Other authors (e.g. Gilson, 2003) have argued that an entrepreneurial culture is necessary to develop a successful VC market. According to Fiedler and Hellmann (2001) the German VC market developed very slowly for a long time because Germans lacked the tolerance for uncertainty necessary for leaving the save harbor of large firm employment. However, the Internet hype and a large number of Germans visiting US business schools have changed their behavior, supporting the German VC market.

121 A very good proxy for entrepreneurship and innovation could be the Community Innovation Surveys (CIS). This survey asks companies in the European Union for their shares of revenue they make with new products. The CIS distinguishes between real inventions and products that were just added to a company's portfolio. The first type is most interesting for my purposes. Unfortunately, a long-term record of the CIS is not available as the first survey started only in 2004.

122 Data for Germany could not be obtained.

Figure 4.12: Founding rates in four European economies (Source: Eurostat)
Data for France was just available from 2003 to 2006.

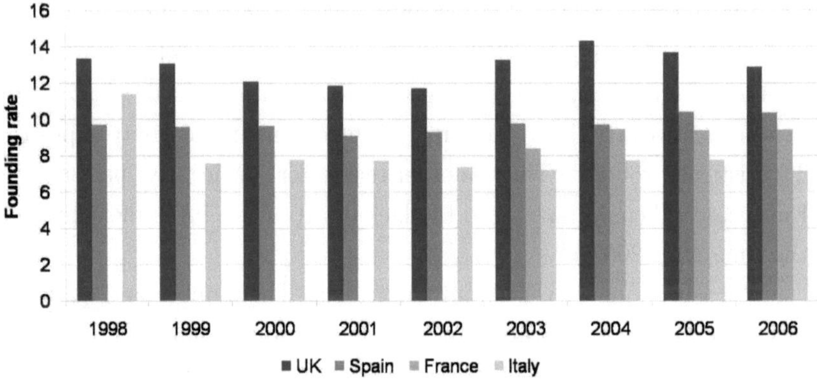

again in 2003. For example, UK - the most important market for VC in Europe - had the highest founding rate in 2004 right at the time when company valuations were low and attractive for VCs to finance portfolio companies.

I proxy the development of innovation over time with the number of patent applications from 27 European countries at the European Patent Office (my data source is Eurostat).[123] This measure has several drawbacks: As Meyer (2008) states, many companies apply for patents but do not want to use them (defensive patenting) or do not apply for a patent because they want to avoid publication of their ideas. Further problems with possibly distorting effects are international patent trading and the fact that innovative business models might not be patentable. Moreover, the number of patents does not consider the different quality of the ideas. An index which is calculated by Pottelsberghe, the chief economist of the European Patent Office, shows that the value of patents was steadily decreasing since the 1980's while the number of patents is growing.[124] Despite the shortcomings in us-

123 Other authors have also used patent applications as a proxy for innovation, e.g., Romain and van Pottelsberghe de la Potterie (2004) and Meyer (2008).

124 The index is called Scope-Year-Index and takes into account for how long and in how many geographies a patent is protected.

Figure 4.13: Indexed development in the number of patent applications in 27 European countries between 1990 and 2005 (Source: Eurostat)

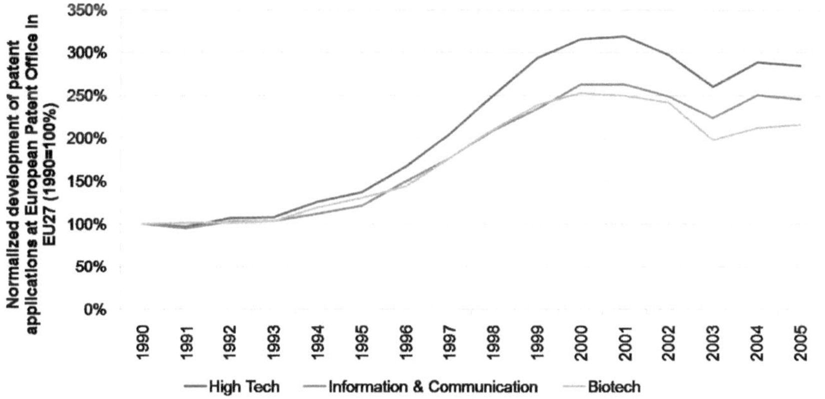

ing the number of patents as measure for innovativeness, it should allow me to differentiate times with less new ideas from times with more.

In a first step I look at the indexed development of the patent applications per year in the sectors "High Tech", "Information and Communication" as well as "Biotechnology" (see Figure 4.13). Obviously, there was a huge increase of these patent types between 1990 and 2005. The level of 2000 is three times the level in 1990. This is not surprising as the three sectors are the core of the "New Economy" that mainly grew at the end of the 1990's with mobile services and the Internet gaining importance. Concerning the development of patent applications in the aftermath of the bubble, I observe only a slight decrease from 2001 to 2003. In 2004 the number of applications increases again. This analysis suggests that the level of innovation in Europe was constantly quite high in the aftermath of the bubble. A general lack of innovation did not cause problems on its' own.

However, a significant disparity between the number of good ideas and the amount of capital in the market might have been an issue: Figure 4.14 compares the indexed accumulated number of patents in the EU27 and the

Figure 4.14: Indexed cumulative development (year 1990: 100%) in the number of patent applications in 27 European countries and the VC fund volume raised between 1990 and 2005 (Source: Eurostat, VentureXpert)

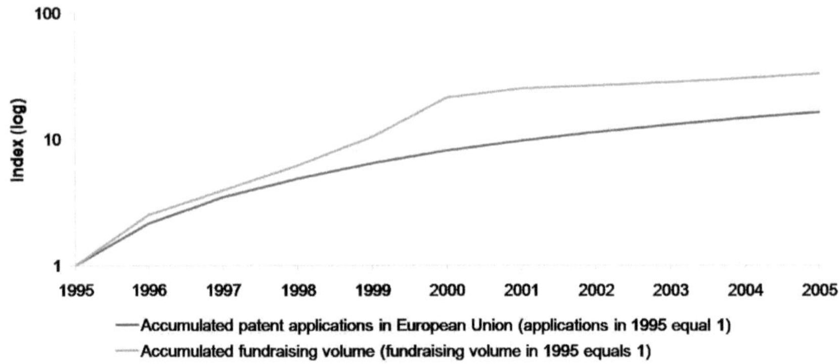

accumulated VC fundraising volume starting in 1995. A significant disparity is obvious. The amount of capital increased many times faster than the number of innovations. Thus, problems of too much money chasing too few (good) deals (Gompers and Lerner, 2000) probably led to price wars.

Figure 4.15 substantiates this suspicion. I display the development of the cumulative fundraising and investment volume in European VC from a VentureXpert extract. I call the difference between both numbers "capital overhang" as it measures how much capital is seeking for investment beyond what is invested directly in a year. I observe, the Internet bubble separates two time periods: Before 1999 the capital overhang is fairly constant at about €2.7B, it peaks in the bubble to about €13B and consolidates at about €9.1B afterwards. The overhang after the bubble is three times the level before. This might have caused that fund managers paid too much for the portfolio companies in the bubble's aftermath (this is corroborated by findings by Da Rin et al., 2006; Meyer, 2008).

The capital overhang-problem I detected for the bubble aftermath seems to confirm the thesis by Da Rin et al. (2006) and Schertler (2007) that in-

Figure 4.15: Development of cumulative committed capital, cumulative investments and the residual cumulative uninvested capital (called "capital overhang") between 1990 and 2008 (Source: VentureXpert)

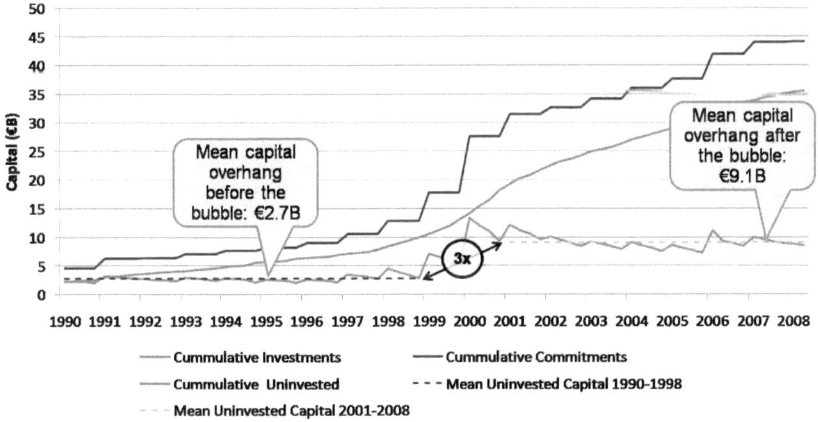

creasing the supply of risk capital does not necessarily promote innovation. They state that increasing Research & Development (R&D) expenditure would be more effective. My data indicate that capital was not the problem - it was more the lack of innovation or the disparity between capital availability and high-quality business opportunities searching for VC funding, which caused troubles.

4.6.3 Limited Partner Behavior

To assess how LP behavior affected the European VC market I investigate two questions:

- How did LPs in Europe time their VC investments?
- Do investors in European VC have the right selection skills?

The first question was partially tackled in the last section. Figure 4.16 further explains how the extensive capital overhang evolved. It was mainly a consequence of the LPs investing behavior: The chart shows the annual fundraising volumes over time. LPs accelerated their fund commitments in 1997 "swamping" the market in the years 1999/2000 with afterpains in the

Figure 4.16: Annual fundraising volumes in Europe between 1990 and 2006 (Source: VentureXpert, Datastream)

years that followed. Surprisingly, despite of the burst of the Internet bubble the volumes of committed capital never dropped below the levels of 1997 in the bubble aftermath. A healthy "downsizing" of the capital searching for investment opportunities never occurred. On the contrary, commitments increased again in the economic recovery 2005/2006. Just in 2007 when capital markets retreated, fund managers had harder times to set up new funds. Overall, the LPs committed capital procyclically. Negative effects on returns may be expected if VC fund managers also invested the capital procyclically into portfolio companies. I explore this in Section 4.6.4.

Looking at the investment skills of LPs in European Venture capital, a conclusive skill assessment is not possible as I lack the necessary data. However, some data points give hints to potential problems in Europe. The first indicator comes from the results I described in Section 4.5. I saw that none of the LP groups in the investigation was able to really set itself apart from the rest and realize attractive positive returns.

The comparison between the LP structure in US and European VC funds reveals another interesting observation (see Figure 4.17): Therefore, I aggregated the LP classes in four groups. The first group consists of Individuals,

Figure 4.17: Share of different LP groups in the fundraising of Venture Capital in the USA and in Europe 2003 (Sources: US data are from NVCA Yearbook 2004, European data are from EVCA/Thomson Reuters/PricewaterhouseCoopers survey)

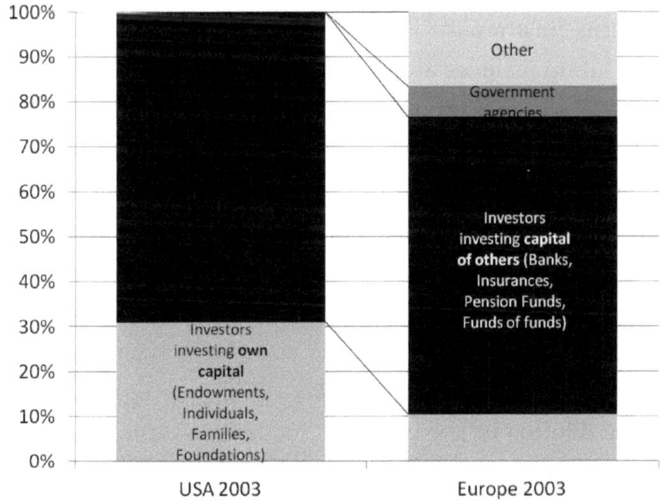

Endowments, Family offices and Foundations. The commonality of these LP types is that they all invest their own capital, i.e., agency problems and other conflicts of interest are not expected to affect their investment decisions.

The second group – consisting of Banks, Asset manager, Insurances, Pension funds and Funds-of-funds – mainly invests capital from others (contributors of capital, policyholders, pensioners and investors). Thus, their investment decision occurs in a principal-agent-relationship. Negative effects on the LPs decisions are possible as the LPs may act opportunistic and maximize their own utility function without deciding in their principals' best interests. LPs from the second group may also have side interests. For example, Banks could invest in VC funds to establish a business relationship, e.g., to sell their consulting work (e.g., for M&As) and underwriting securities for equity issues (e.g., Hellmann et al., 2008).

The third group are government agencies with other interests than just investment returns by investing in VC funds. For example, they want to support certain regions or innovation and entrepreneurship in a country in general (Lerner et al., 2007). Thus, they may consider the creation of investment returns "just a positive side effect".

The fourth group includes all other LPs.

As Figure 4.17 shows, European and US VC funds have a quite different LP structure. The share of investors investing other people's assets is quite similar at about 70%. However, the remaining 30% of capital is from Individuals, Endowments, Family offices and Foundations in the US, while this group contributes just 10% in Europe. The remaining 20% in Europe are from Government agencies (7%) and other LP classes. This result shows that agency problems and side interests can play a significantly more important role in Europe than in the USA.

I expect the dilution of the investors' focus on returns to be mainly due to the lack of Endowments in Europe. As Lerner et al. (2007) found, US endowments have proven superior fund selection skills compared to other investors in the 1990. They are supposed to proactively use the private information from a fund for future investment decisions and react better to market changes than others. Endowments received a lot of attention in the press and from academia. Other investor groups tried to adopt endowments' investment strategies and patterns (Hobohm, 2008a). Although Hobohm (2008b) does not find superior performance for Endowments after 1998 (in line with my results) they generally seem to have a good fit to the venture capital market. Their investment sizes correspond quite well to the fund sizes of VC funds and they are more entrepreneurial than other investors as they invest their own capital.

In conclusion, the lack of a broad basis of investors who just focus on generating high returns for themselves probably induced a low average fund (manager) quality in Europe.

Figure 4.18: Timing of VC Investments between 1997 and 2009 vs. the development of the MSCI Europe (Source: EVCA, Datastream)

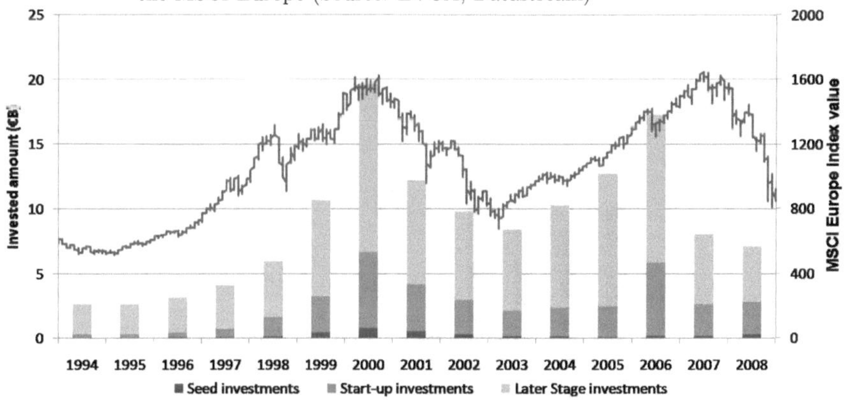

4.6.4 Fund Manager Behavior

Successful investing in young companies requires a "nose" for industry trends and good market timing skills. I cannot evaluate if VC fund managers saw the right trends in the industries they invested in or if they overslept them until it was too late to buy into them before they picked up. This would require more intimate data on the ventures which applied for capital from VC funds, but did not get financing. However, I can assess, if fund managers timed the general market environment reasonably well. Nowak et al. (2004) found out that - especially for VC funds - the investment timing is very important. Good timing in the authors' understanding means, fund managers should invest in periods of low valuation and divest in high-valuation times. A good investment timing positively affects the returns the fund managers generate with their portfolio companies.

I perform two analyzes to explore the fund managers' investment timing: In Figure 4.18 I look at the joint movement between the invested amounts in different VC stages over time and the MSCI Europe. Obviously, the fund managers' investment pattern is as procyclical as the LPs' commitment pat-

Figure 4.19: Average number of first-round investments per active fund between 1994 and 2008 vs. the development of the MSCI Europe (Source: CEPRES, Datastream)

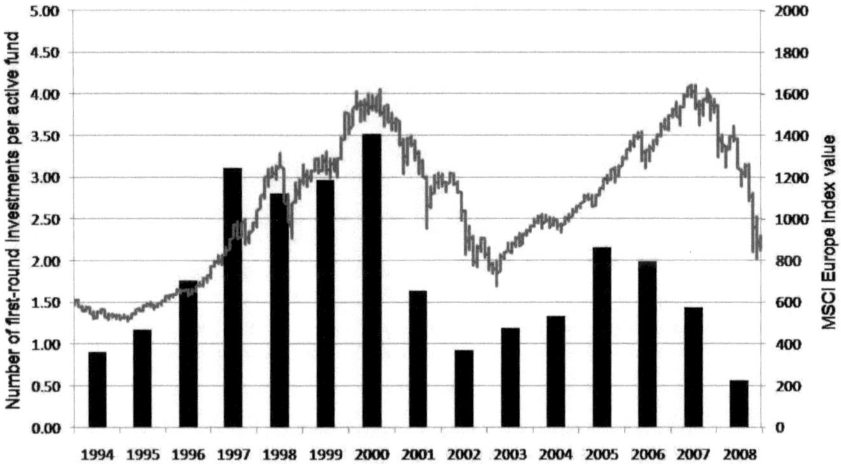

tern. During times when stock market valuations are high, GPs spent most on portfolio companies and vice versa.

One could argue that this behavior is rational as far as fund managers who start investing into portfolio companies in low-valuation times usually commit only little amounts to kick-start the portfolio companies' businesses. When the companies are growing greater amounts are usually invested during further financing rounds. Especially during times of high market growth and high valuation, the managers need to spend a lot to make their portfolio companies fit for a sale to the capital market or another buyer. However, this argumentation would mean that fund managers should continuously finance good new business ideas – especially during times of low valuation shortly before an economic recovery. Accordingly, they could participate in the following economic growth period. Therefore, I investigate when fund managers conducted first-round commitments to their portfolio companies over time.

Total numbers of first-round investments do not say much in this respect as the number of fund managers searching for investment opportunities varies significantly over time. However, a good proxy for the timing behavior of fund managers is how many first-round investments were conducted on average per active fund manager in the market over time.

My calculations are based on an extended *CEPRES* sample. Beyond the data on fully and partially realized investments which were presented in Section 4.3, I also received statistics on the number of unrealized portfolio companies in the funds and the years these portfolio companies had their first financing round. This information is necessary as distortions could arise neglecting unrealized investments - especially in the last few years 2007-09.

Based on this information I am able to estimate an average share of the funds which spend capital in their first, second, third and so on fund year. Unsurprisingly this share starts with 100% in the first year and decreases over time to 6% in the eighth year. Because I know the funds' vintage years in my sample, the shares allow me to estimate how many funds are searching for investment opportunities in a specific year.

Finally, the ratio between the number of first-round investments in a year and the number of funds searching for investment opportunities equals the average investment starts per fund over time. As Figure 4.19 shows, fund managers also acted procyclically concerning the first-round investments. The managers didn't invest continuously, but significantly reduced the number of companies they added to their portfolios - especially in the years 2002-2004. This may partially explain why I saw comparatively few IPOs in 2005 to 2007. Fund managers simply hadn't funded sufficient portfolio companies to support them to an IPO.

4.7 Conclusions

The insight that the Venture Capital market develops in waves is not new (e.g., Smith and van den Berg, 2006). Years of good investment returns alternate with years of bad returns because market conditions are not always in favor for fund managers and their portfolio companies. Remarkably, however, is the fact that VC returns in Europe were not attractive in the whole last decade. And even more surprising is the fact, that fund managers rather burned a lot of capital and none of the investor groups in VC were really immune to this problem.

Interestingly, the bad performance of the funds in Europe seems not to be due to the macroeconomic environment that didn't offer attractive times for investing in portfolio companies, building and exiting them via IPO or trade sale. A business cycle with good opportunities for fund managers was available in the time span between 2002 and 2008. This is confirmed by the (few) portfolio companies with good returns which were supported in 2003/2004. However, these opportunities were hardly exploited. Fund managers preferred investing procyclically and spent heavily in times of high valuations while hardly financing portfolio companies during low valuation periods.

A potential reason for this disadvantageous investment behavior of GPs could be that the procyclical investment style of LPs, amplified by the flooding of capital in the market, led to a significant dilution of the average fund manager quality.[125] This hypothesis is in line with a theoretical model by Smith and van den Berg (2006). Moreover, an increased share of unsophisticated fund managers, who invested unsuccessfully and who also derogated the skilled fund managers' investment scope, apparently also induced an overall negative picture of the European VC market.

A question is how the European VC market can overcome the inevitable escape of investors from this low quality market. My answer is, it shouldn't! I predict that not every fund managers will survive this market down-sizing, but this will be healthy for the industry overall.[126] If and when investors leave the market, an improvement in the quality can be expected both on the fund manager and the portfolio company level. The average performance will recover to attractive levels and European venture capital will reclaim the function it should have: A capital market with high returns for

125 In my full fund sample I find that some fund managers consistently realized good positive returns over time even with funds of the vintage years 1999 and 2000.

126 For a similar suggestion for the US VC market see also Kedrosky (2009).

investors which supports high quality start-ups with the ability to compete internationally and which secures the strength of the European economy for the future.

Bibliography

Aizenman, Joshua / Kendall, Jake (2008): The Internationalization of Venture Capital and Private Equity, *SSRN Working Paper*, URL `http://ssrn.com/abstract=1275544`.

Akantziliotou, K. / Rigby, R. A. / Stasinopoulos, D. M. (2002): The R Implementation of Generalized Additive Models for Location, Scale and Shape, in: Statistical Modelling in Society: Proceedings of the 17th International Workshop on Statistical Modelling, edited by Stasinopoulos, D. M. and Touloumi, G., Greece 2002, pp. 75–83.

Akerlof, George A. (1970): The Market for 'Lemons': Quality Uncertainty and the Market Mechanism, in: *Quarterly Journal of Economics*, Vol. 84, pp. 488–500.

Albano, Gian Luigi / Lizzeri, Alessandro (2001): Strategic Certification and Provision of Quality, in: *International Economic Review*, Vol. 42, pp. 267–283.

Askar, Atif (2008): Performance-Messung bei Venture Capital-Fonds, Marburg 2008, ISBN 3828895166.

Balboa, Marina Ramón / Martí, José Pellón (2007): Factors That Determine the Reputation of Private Equity Managers in Developing Markets, in: *Journal of Business Venturing*, Vol. 22, pp. 453–480.

Barry, Christopher B. (1994): New Directions in Research On Venture Capital Finance, in: *Financial Management*, Vol. 23, pp. 3–15.

Beauchamp, Charles F. (2006): An Empirical Examination of 'Learning' Within the Private Equity Industry, *Mississippi State University Working Paper.*

Black, B. / Gilson, R. (1998): Venture Capital and the Structure of Capital Markets: Banks versus Stock Markets, in: *Journal of Financial Economics*, Vol. 47, pp. 243–278.

Bonini, Stefano / Alkan Aktuccar, Senem (2009): The Macro and Political Determinants of Venture Capital Investments Around the World, *SSRN Working Paper*, URL `http://ssrn.com/abstract=945312`.

Bottazzi, Laura / Da Rin, Marco (2002): Venture Capital in Europe and the Financing of Innovative Companies, in: *Economic Policy*, Vol. 17, pp. 229–269.

Box, George E. P. / Cox, David R. (1964): An Analysis of Transformation, in: *Journal of the Royal Statistical Society*, Vol. 26, pp. 211–246.

Brigham, Eugene F. / Gapenski, Louis C. (1997): Financial Management: Theory and Practice, 8. ed., Fort Worth 1997, ISBN 9780030177897.

Buchner, Axel / Kaserer, Christoph / Wagner, Niklas (2008): A Stochastic Model for the Cash Flow Dynamics of Private Equity Funds – Theory and Empirical Evidence, *Working Paper Technical University Munich (TUM).*

Buchner, Axel / Krohmer, Philipp / Schmidt, Daniel / Wahrenburg, Mark (2010): Projection of Venture Capital Fund Performance: A Simulation Approach, in: Private Equity: Fund Types, Risks and Returns, and Regulation, edited by Cumming, Douglas J., Hoboken (New Jersey) 2010, pp. 197–228.

BVCA (2008): BVCA Private Equity and Venture Capital Performance Measurement Survey 2008, London 2008.

Chan, Yuk-Shee (1983): On the Positive Role of Financial Intermediation in Allocation of Venture Capital in a Market with Imperfect Information, in: *The Journal of Finance*, Vol. 38, pp. 1543–1568.

Chen, P. / Baierl, G. / Kaplan, Steven N. (2002): Venture Capital and Its Role in Strategic Asset Allocation, in: *Journal of Portfolio Management*, Vol. 28, pp. 83–90.

Cochrane, John H. (2005): The Risk and Return of Venture Capital, in: *Journal of Financial Economics*, Vol. 75, pp. 3–52.

Cook, P. (1998): Private Equity Investing, in: Alternative Investment Strategies, edited by Jaffer, Sohail, London 1998, pp. 13–27, ISBN 1855646366.

Cooper, Alan (2006): Selecting and Working With a Placement Agent, in: A Guide to Private Equity Fund Placement Specialists, edited by Private Equity International, 2006, pp. 25–28, ISBN 9781904696148.

Cumming, Douglas / Walz, Uwe (2004): Private Equity Returns and Disclosure Around the World, in: *EFA 2004 Maastricht, EEA 2004 Madrid, EFM 2005 Madrid, Journal of International Business Studies, forthcoming.*

Cumming, Douglas J. / MacIntosh, Jeffrey G. (2001): Venture Capital Investment Duration in Canada and the United States, in: *Journal of Multinational Financial Management*, Vol. 11, pp. 445–463.

Cumming, Douglas J. (2006): The Determinants of Venture Capital Portfolio Size: Empirical Evidence, in: *Journal of Business*, Vol. 79, pp. 1083–1126.

Da Rin, M. / Nicodano, G. / Sembenelli, A. (2006): Public Policy and the Creation of Active Venture Capital Markets, in: *Journal of Public Economics*, Vol. 90, pp. 8–9.

Diller, Christian (2007): Private Equity: Rendite, Risiko und Markteinflussfaktoren: Eine empirische Analyse europäischer Private-Equity-Fonds, 1. Aufl., Bad Soden / Ts. 2007, ISBN 9783933207586.

Dorsey, Terry (2000): A Portfolio Model for Venture Capital Performance Measurement and Investment Selection, *GARP Working Paper*, URL `http://www.garp.com/documents/Articles/Dorsey.pdf`.

Driessen, Joost / Tse-Chun, Lin / Phalippou, Ludovic (2009): A New Method to Estimate Risk and Return of Non-Traded Assets from Cash Flows: The Case of Private Equity Funds, *NBER Working Paper No. W14144*.

Eilers, Paul H. / Marx, Brian D. (1996): Flexible Smoothing With B-Splines and Penalties, in: *Statistical Science*, Vol. 11, pp. 89–121.

Fenn, George W. / Liang, Nellie / Prowse, Stephen (1995): The Economics of the Private Equity Market, *Board of Governors of the Federal Reserve System*, URL `www.federalreserve.gov/pubs/staffstudies/1990-99/ss168.pdf`.

Fiedler, Marc-Oliver / Hellmann, Thomas (2001): Against All Odds: The Late But Rapid Development of the German Venture Capital Industry, in: *Journal of Private Equity*, Vol. 4, pp. 31–45.

Gilson, Ronald J. (2003): Engineering a Venture Capital Market: Lessons from the American Experience, in: *Stanford Law Review*, Vol. 55, pp. 1067–1092.

Giot, P. / Schwienbacher, A. (2005): IPOs, Trade Sales and Liquidations: Modelling Venture Capital Exits Using Survival Analysis, *Economics Working Paper No 0312006. Archive at WUSTL.*

Golec, Joseph H. (1992): Empirical Tests of a Principal-Agent Model of the Investor-Investment Advisor Relationship, in: *The Journal of Financial and Quantitative Analysis*, Vol. 27, pp. 81–95.

Gompers, Paul / Lerner, Josh (1999): An Analysis of Compensation in the U.S. Venture Capital Partnership, in: *Journal of Financial Economics*, Vol. 51, pp. 3–44.

Gompers, Paul A. / Lerner, Josh / Blair, Margaret M. / Hellmann, Thomas (1998): What Drives Venture Capital Fundraising?, in: *Brookings Papers on Economic Activity: Microeconomics*, Vol. 1998, pp. 149–204.

Gompers, Paul A. / Lerner, Josh (1999): What Drives Venture Capital Fundraising?, *NBER Working Paper No. 6906.*

Gompers, Paul Alan / Lerner, Joshua (1997): Risk and Reward in Private Equity Investments: The Challenge of Performance Assessment, in: *Journal of Private Equity*, Vol. 1, pp. 5–12.

Gompers, Paul Alan / Lerner, Josh (2000): Money Chasing Deals? The Impact of Fund Inflows On Private Equity Valuations, in: *Journal of Financial Economics*, Vol. 50, pp. 281–325.

Gompers, Paul Alan / Lerner, Joshua (2001): The Money of Invention: How Venture Capital Creates New Wealth, Boston (Mass.) 2001, ISBN 157851326X.

Gompers, Paul Alan / Lerner, Joshua (2004): The Venture Capital Cycle, 2. ed., Cambridge (Mass.) 2004, URL `http://www.loc.gov/catdir/toc/fy052/2004040286.html`, ISBN 0262072556.

Gompers, Paul Alan / Kovner, Anna / Lerner, Joshua / Scharfstein, D. (2005): Venture Capital Investment Cycles: The Impact of Public Markets, *NBER Working Paper No. W11385*, URL `http://citeseerx.ist.psu.edu/viewdoc/download?doi=10.1.1.61.3638&rep=rep1&type=pdf`.

Gormann, M. / Salmann, W. A. (1989): What Do Venture Capitalists Do?, in: *Journal of Business Venturing*, Vol. 4, pp. 231–248.

Gottschalg, Oliver F. / Phalippou, Ludovic / Zollo, M. (2004): Performance of Private Equity Funds: Another Puzzle?, *INSEAD Working Paper*.

Gottschalg, Oliver F. / Phalippou, Ludovic (2008): The Performance of Private Equity Funds, in: *The Review of Financial Studies*, Vol. 22, pp. 1747–1776.

Grabenwarter, Ulrich / Weidig, Tom (2005): Exposed to the J-Curve: Understanding and Managing Private Equity Fund Investments, London 2005, ISBN 1843741490.

Gras, Isabelle (2003): The Power to Rate: Eine Untersuchung zur Rolle der Ratingagenturen auf den internationalen Finanzmärkten, in: *REGEM Analysis*, Vol. 6, pp. 1–40.

Groh, Alexander Peter / Liechtenstein, Heinrich (2009): The First Step of the Capital Flow from Institutions to Entrepreneurs: The Criteria for Sorting Venture Capital Funds, *IESE Business School Working Paper No. 795*.

Grossman, Sanford J. / Hart, Oliver D. (1983): An Analysis of the Principal-Agent Problem, in: *Econometrica*, Vol. 51, pp. 7–46.

Guliner, Mathew (2007): Capital Raising - Placement Agents, *Young Venture Capital Society Working Paper*, URL `http://www.yvcs.org/uploads/1189810728PlacementAgent.pdf`.

Hagenmüller, Moritz (2004): Entrepreneurial and Financial Studies. Vol. 5: Investor Relations von Private Equity Partnerships, Sternenfels 2004, ISBN 3896732390.

Harhoff, Dietmar / Licht, Georg / Beise, Marian / Felder, Johannes / Nerlinger, Eric / Stahl, Harald (1996): Innovationsaktivitäten kleiner und mittlerer Unternehmen - Ergebnisse des Mannheimer Innovationspanels, *Baden-Baden, Nomos*.

Harris, Clint (2003): The Challenging Role of a 'Gatekeeper' in the Rapidly Changing World of Private Equity, in: *Israel Venture Capital Journal*, Vol. September, pp. 6–7, URL `http://www.grovestreetadvisors.com/news/2003.9_IVCJ.pdf`.

Hastie, T. J. / Tibshirani, R. J. (1993): Varying Coefficient Models (With Discussion), in: *Journal of the Royal Statistical Society*, Vol. 55, pp. 757–796.

Hege, Ulrich / Palomino, Frédéric / Schwienbacher, Armin (2009): Venture Capital Performance: The Disparity Between Europe and the United States, in: *Revue de L'Association FranÃ§aise de Finance*, Vol. 30, URL `https://studies2.hec.fr/jahia/webdav/site/hec/shared/sites/hege/acces_anonyme/papers/HegePalominoSchwienbacher_VCPerf2009.pdf`.

Heinke, Volker (2000): Der Signal- und Zertifizierungswert von Credit Ratings am Euromarkt, in: *Die Betriebswirtschaft*, Vol. 60, pp. 314–335.

Hellmann, Thomas / Lindsey, L. / Puri, M. (2008): Building Relationships Early: Banks in Venture Capital, in: *Review of Financial Studies*, Vol. 21, pp. 513–541.

Hillebrand, Ernst (2001): Schlüsselstellung im globalisierten Kapitalismus - Der Einfluss privater Rating-Agenturen auf Finanzmärkte und Politik, in: Die Privatisierung der Weltpolitik: Entstaatlichung und Kommerzialisierung im Globalisierungsprozess, edited by Brühl, Tanja Vol. 11, Bonn 2001, pp. 150–171, ISBN 3801202992.

Hobohm, Daniel (2008): Investors in Private Equity Funds: Fund Preferences, Market Inefficiencies & Flexible Endowments, *LMU Working Paper*, URL `http://ssrn.com/abstract=1298589`.

Hobohm, Daniel (2008): Investors in Private Equity Funds: Large-Scale Performance Analysis and the Question If Location Matters, *LMU Working Paper*, URL `http://ssrn.com/abstract=1286139`.

Holmstrom, Bengt (1979): Moral Hazard and Observerbility, in: *Bell Journal of Economics*, Vol. 10, pp. 74–91.

Ick, Matthias M. (2005): Performance Measurement and Appraisal of Private Equity Investments Relative to Public Equity Markets, *CEPRES Working Paper*, URL `http://ssrn.com/abstract=871931`.

Jean, William H. (1970): The Analytical Theory of Finance: A Study of the Investment Decision Process of the Individual and the Firm, New York 1970, ISBN 0030820855.

Jeng, L. A. / Wells, P. C. (2000): The Determinants of Venture Capital Funding: Evidence Across Countries, in: *Journal of Corporate Finance*, Vol. 6, pp. 241–289.

Jones, C. M. / Rhodes-Kropf, M. (2003): The Price of Diversifiable Risk in Venture Capital and Private Equity, *AFA 2003 Washington, DC Meetings*, URL `http://ssrn.com/abstract=342841`.

Kanniainen, Vesa / Keuschnigg, Christian (2003): The Optimal Portfolio of Start-Up Firms in Venture Capital Finance, in: *Journal of Corporate Finance*, Vol. 9, pp. 521–534.

Kaplan, Steven N. / Schoar, Antoinette (2005): Private Equity Performance: Returns, Persistence, and Capital Flows, in: *Journal of Finance*, Vol. 60, pp. 1791–1823.

Kaserer, Christoph / Diller, Christian (2004): European Private Equity Funds - A Cash Flow Based Performance Analysis, *CEFS Working Paper*.

Kaserer, Christoph / Diller, Christian (2007): What Drives Private Equity Returns? – Fund Inflows, Skilled GPs, And/Or Risk?, *CEFS Working Paper*, URL `http://ssrn.com/abstract=665602`.

Kedrosky, Paul (2009): Right-Sizing the U.S. Venture Capital Industry, *Kauffman Foundation Small Research Projects Research*, URL `http://ssrn.com/abstract=1456431`.

Koenker, Roger / Hallock, Kevin F. (2001): Quantile Regression, in: *Journal of Economic Perspectives*, Vol. 15, pp. 143–156.

Kofman, Fred / Lawarree, Jacques (1993): Collusion in Hierarchical Agency, in: *Econometrica*, Vol. 61, pp. 629–656, URL `http://ideas.repec.org/a/ecm/emetrp/v61y1993i3p629-56.html`.

Kortum, S. / Lerner, Josh (2000): Assessing the Contribution of Venture Capital to Innovation, in: *The RAND Journal of Economics*, Vol. 31, pp. 674–692.

Krohmer, Philipp / Man, Kwok-Sing (2008): Modeling Default Risk of Private Equity Funds - A Market-Based Framework, in: *CEPRES Working Paper*.

Laine, M. / Torstila, S. (2005): The Exit Rates of Liquidated Venture Capital Funds, in: *Journal of Entrepreneurial Finance and Business Ventures*, Vol. 10, pp. 53–73.

Leland, Hayne / Pyle, David (1977): Information Asymmetries, Financial Structure, and Financial Intermediation, in: *The Journal of Finance*, Vol. 32, pp. 371–387.

Lerner, Josh / Schoar, Antoinette (2004): The Illiquidity Puzzle: Theory and Evidence from Private Equity, in: *Journal of Financial Economics*, Vol. 72, pp. 3–40.

Lerner, Josh (2005): A Note On the Private Equity Fundraising Process, in: Venture Capital and Private Equity: A Casebook, edited by Lerner, Josh and Hardymon, Felda. and Leamon, Ann., Hoboken, NJ 2005, pp. 105–110, ISBN 0471230693.

Lerner, Josh / Hardymon, Felda. / Leamon, Ann. (2005): Venture Capital and Private Equity: A Casebook, 3. ed., Hoboken, NJ 2005, ISBN 0471230693.

Lerner, Josh / Schoar, Antoinette / Wongsunwai, Wan (2007): Smart Institutions, Foolish Choices: The Limited Partner Performance Puzzle, in: *Journal of Finance*, Vol. 62, pp. 731–764.

Liu, Wei-Lin (2005): Motivating and Compensating Investment Advisors, in: *Journal of Business*, Vol. 78, pp. 2317–2349.

Lizzeri, Alessandro (1999): Information Revelation and Certification Intermediaries, in: *RAND Journal of Economics*, Vol. 30, pp. 214–231.

Ljungqvist, Alexander / Richardson, Matthew (2003): The Investment Behaviour of Private Equity Fund Managers, *RICAFE Working Paper No. 005*, URL `http://www.lse.ac.uk/collections/RICAFE/pdf/RICAFE-WP05-Ljungqvist.pdf`.

Lossen, Ulrich (2006): The Performance of Private Equity Funds: Does Diversification Matter?: Discussion Paper No. 192, *Munich Business Research Working Paper Series No. 2006-14*.

Malherbe, Etienne de (2005): Modelling Private Fquity Funds and Private Equity Collateralised Fund Obligations, in: *The Journal of Alternative Investments*.

Marco, Jose Antonio (2002): Winning Strategies for Private Equities in Spain – Learning from the US, *Harward Working Paper*, URL `http://www.avco.at/upload/medialibrary/17094_0_StrategiesPESpain.pdf`.

Martí, José Pellón / Balboa, Marina Ramón (2003): Characterisation of the Reputation of Private Equity Managers: Evidence in Spain, in: *EFMA 2003 Helsinki Meetings*, URL `http://ssrn.com/abstract=391662`.

Mayer, Colin / Schoorsb, Koen / Yafehc, Yishay (2005): Sources of Funds and Investment Activities of Venture Capital Funds: Evidence from Germany, Israel, Japan and the United Kingdom, in: *Journal of Corporate Finance*, Vol. 11, pp. 586–608.

McDaniel, William R. / McCarty, Daniel E. / Jessell, Kenneth A. (1988): Discounted Cash Flow with Explicit Reinvestment Rates: Tutorial and Extension, in: *The Financial Review*, Vol. 23, pp. 369–385.

Meuleman, Miguel / Manigart, Sophie / Lockett, Andy / Wright, Mike (2006): Transaction Costs, Behavioral Uncertainty and the Formation of Interfirm Cooperations: Syndication in the UK Private Equity Market, *Universiteit Gent Working Paper No. 2006/359*, URL `http://www.FEB.UGent.be/nl/Ondz/wp/Papers/wp_06_359.pdf`.

Meyer, Colin / Schoors, Koen / Yafeh, Yishay (2005): Sources of Funds and Investment Activities of Venture Capital Funds - Evidence from Germany, Isreal, Japan and United Kingdom, in: *Journal of Corporate Finance*, Vol. 11, pp. 586–608.

Meyer, T. / Weidig, Tom (2003): Modelling Venture Capital Funds, in: *Risk Magazine*, Vol. 16, pp. 89–92.

Meyer, Thomas (2006): Venture Capital in Europe: Spice for European Economies, in: *Economics (Deutsche Bank Research)*, Vol. 60, URL `http://www.dbresearch.de/PROD/CIB_INTERNET_EN-PROD/PROD0000000000202800.pdf`.

Meyer, Thomas (2007): The Public Sector's Role in the Promotion of Venture Capital Markets, *SSRN Working Paper*, URL `http://ssrn.com/abstract=1019988`.

Meyer, Thomas (2008): Venture Capital: Brücke zwischen Idee und Innovation?, in: *Economics (Deutsche Bank Research)*, Vol. 65.

Mills, Lauren (2005): The New Road, in: *Real Deals*, Vol. January, pp. 19–24.

Moskowitz, Tobias J. / Vissing-Jorgensen, Annette (2002): The Returns to Entrepreneurial Investment: A Private Equity Premium Puzzle?, in: *The American Economic Review*, Vol. 92, pp. 745–778.

Nelder, J. A. / Wedderburn, R. W. M. (1972): Generalized Linear Models, in: *Journal of the Royal Statistical Society*, Vol. 135, pp. 370–384.

Nowak, E. / Knigge, A. / Schmidt, Daniel (2004): On the Performance of Private Equity Investments: Does Market Timing Matter?, *EFMA 2004 Basel Meetings Paper*, URL http://ssrn.com/abstract=492982.

Partnoy, Frank (1999): The Siskel and Ebert of Financial Markets?: Two Thumps Down for the Credit Rating Agencies, in: *Washington University Law Quarterly*, Vol. 77, pp. 619–714.

Peng, L. (2001): Building a Venture Capital Index, *Yale ICF Working Paper No. 00-51.*

Peyrache, Eloic / Quesada, Lucia (2004): Strategic Certification, *University of Wisconsin Madison Working Paper*, URL https://studies2.hec.fr/jahia/webdav/site/hec/shared/sites/Peyrache/acces_anonyme/research/strategic.pdf.

Phalippou, Ludovic / Zollo, M. (2005): The Performance of Private Equity Funds, *Warton Working Paper*, URL http://fic.wharton.upenn.edu/fic/papers/05/0542.pdf.

Phalippou, Ludovic / Gottschalg, Oliver F. (2009): The Performance of Private Equity Funds, in: *The Review of Financial Studies*, Vol. 22, pp. 1747–1776.

Private Equity International (2003): Placement Power, in: *Private Equity International*, Vol. July/August, pp. 23–24, URL `http://www.mvision.com/PDFs/12.pdf`.

Private Equity International (**2006**): A Guide to Private Equity Fund Placement Specialists, 2nd edition, 2006, ISBN 9781904696148.

Prowse, Stephen D. (1998): The Economics of the Private Equity Market, in: *Federal Reserve Bank of Dallas – Economic Review*, Vol. Third Quarter, pp. 21–34, URL `http://economics.sas.upenn.edu/~rafaels/doc/foundation.pdf`.

Quigley, J. / Woodward, S. E. (2003): An Index for Venture Capital, *University of California at Berkeley Working Paper No. E03-333*, URL `http://repositories.cdlib.org/cgi/viewcontent.cgi?article=1059&context=iber/econ`.

Rigby, R. A. / Stasinopoulos, D. M. (2001): The GAMLSS Project: A Flexible Approach to Statistical Modelling, in: New Trends in Statistical Modelling: Proceedings of the 16th International Workshop on Statistical Modelling, edited by Klein, B. and Korsholm, L., Odense, Denmark 2001, pp. 249–256.

Rigby, R. A. / Stasinopoulos, D. M. (2004): Smooth Centile Curves for Skew and Kurtotic Data Modelled Using Box-Cox Power Exponential Distribution, in: *Stat Med.*, Vol. 23, pp. 3053–76.

Rigby, R. A. / Stasinopoulos, D. M. (2005): Generalized Additive Models for Location, Scale and Shape, (With Discussion), in: *Applied Statistics*, Vol. 54, pp. 507–554.

Romain, Astrid / Potterie, Bruno van Pottelsberghe de la (2004): The Determinants of Venture Capital: A Panel Data Analysis of 16 OECD Countries, *Working Paper WP-CEB 04/015. Université Libre de Bruxelles*.

Rosenbaum, Jens (2004): Der Einsatz von Rating-Agenturen zur Kapitalmarktregulierung in den USA: Ursachen und Konsequenzen, in: *REGEM Analysis*, Vol. 8, URL http://www.chinapolitik.de/studien/regem/regem_no8.pdf.

Ross, Stephen A. (1977): The Determination of Financial Structure: The Incentive Signaling Approach, in: *Bell Journal of Economics*, Vol. 8, pp. 23–40.

Sahlman, William A. (1990): The Structure and Governance of Venture Capital Organizations, in: *Journal of Financial Economics*, Vol. 27, pp. 473–521.

Schertler, Andrea (2007): Knowledge Capital and Venture Capital Investments: New Evidence from European Panel Data, in: *German Economic Review*, Vol. 8, pp. 64–88.

Schmidt, Daniel / Wahrenburg, Mark (2004): Contractual Relations Between European VC–Funds and Investors: The Impact of Reputation and Bargaining Power On Contractual Design, *RICAFE Working Paper No. 008*.

Schwienbacher, Armin (2005): An Empirical Analysis of Venture Capital Exits in Europe and the United States, *EFA 2002 Berlin Meetings Discussion Paper*, URL http://ssrn.com/abstract=302001.

Smith, Han T. J. / Berg, Ward A. van den (2006): Private Equity Waves, in: *Tinbergen Institute Discussion Papers No. 06-053/2*.

Smith, Roy / Walter, Ingo (2001): Rating Agencies: Is There an Agency Issue?, *SSRN eLibrary*, URL http://www1.worldbank.org/finance/html/credit_report_references.html.

Sorrentino, Theresa (2003): Note on Private Placement Memoranda, *Tuck School of Business at Dartmouth College*, URL http://mba.tuck.dartmouth.edu/pecenter/research/pdfs/Private_placement_memo.pdf.

Spence, Michael (1973): Job Market Signaling, in: *Quarterly Journal of Economics*, Vol. 87, pp. 355–374.

Steiner, Manfred / Heinke, Volker G. (1996): Rating aus Sicht der modernen Finanzierungstheorie, in: Handbuch Rating, edited by Büschgen, Hans E. and Everling, Oliver, Wiesbaden 1996, pp. 579–b628.

Stiglitz, Joseph E. (1975): The Theory of Screening, Education and the Distribution of Income, in: *American Economic Review*, Vol. 65, pp. 283–300.

Takahashi, Dean / Alexander, Seth (2002): Illiquid Alternative Asset Fund Modeling, in: *Journal of Portfolio Management*, Vol. 28, pp. 90–100.

Tobin, James (1958): Estimation for Relationships With Limited Dependent Variables, in: *Econometrica*, Vol. 26, pp. 24–36.

Tykvova, Tereza (2007): Who Chooses Whom? Syndication, Skills and Reputation, in: *Review of Financial Economics*, Vol. 16, pp. 5–28.

Wehrspohn, Uwe (2002): Serie: Bestimmung von Ausfallwahrscheinlichkeiten - Teil 4: Ausfallwahrscheinlichkeiten im Konjunkturzyklus - Credit Portfolio View, in: *RiskNews*, Vol. 11, pp. 45–64.

Wehrspohn, Uwe (2003): Serie: Bestimmung von Ausfallwahrscheinlichkeiten - Teil 5: Integrative Modelle - Credit Risk Evaluation Model, in: *RiskNews*, Vol. 1, pp. 1–14.

Weidig, Tom (2002): Risk Model for Venture Capital Fund, *QuantExperts Working Paper*, URL http://ssrn.com/abstract=365881.

Weidig, Tom (2002): Towards a Risk Model for Venture Capital Funds: Liquidity and Performance Forecasting, *QuantExperts Working Paper*, URL www.ssrn.com/author=336052.

Weidig, Tom / Mathonet, Pierre-Yves (2004): The Risk Profiles of Private Equity, *SSRN Working Paper*, URL http://ssrn.com/abstract=495482.

White, Lawrence (2001): The Credit Rating Industry: An Industrial Organization Analysis, *Worldbank Working Paper*, URL http://www1.worldbank.org/finance/html/credit_report_references.html.

Woodward, Susan E. / Hall, Robert E. (2003): Benchmarking the Returns to Venture, *Working Paper*, URL http://ssrn.com/abstract=474181.

Woodward, Susan E. (2009): Measuring Risk for Venture Capital and Private Equity Portfolios, *Sand Hill Econometrics Working Paper*, URL http://ssrn.com/abstract=1458050.

Wright, Mike / Lockett, Andy (2003): The Structure and Management of Alliances: Syndication in the Venture Capital Industry, in: *Journal of Management Studies*, Vol. 40, pp. 2073–2102.

Yung, Chris (2008): The Long-Run Supply and Demand for Venture Capital Funds: Information and Endogenous Entry, *SSRN Working Paper*, URL http://ssrn.com/abstract=1356906.

Appendices for Chapter 2

1 Proof of Lemma 1

Proof. The average reputation after certifying a PEF has to be positive for the certifier to remain in business, i. e. the reputation change from investors $x \cdot R_{I,g} - y \cdot R_{I,b}$ has to be geater or equal to 0. In the long run, the certifier wants to keep his reputation for any x and y. In consequence

$$\begin{aligned} x \cdot R_{I,g} - y \cdot R_{I,b} &\geq 0 \\ x \cdot R_{I,g} &\geq y \cdot R_{I,b} \end{aligned}$$

With $R_{I,b} \geq R_{I,g}$ and $x > 0$ the inequality returns

$$\begin{aligned} 1 \geq \frac{R_{I,g}}{R_{I,b}} &\geq \frac{y}{x} \\ x &\geq y\,. \end{aligned}$$

For $x = 0$ we get $-y \cdot R_{I,b} \geq 0$, so y would have to be 0. In this case the certifier would not issue any certificates. □

2 Proof of Lemma 2

Proof. If a certifier blindly certifies with $z > 0$ funds his reputation gain will be $z \cdot \gamma \cdot R_{I,g} - z \cdot (1 - \gamma) \cdot R_{I,b}$ which has to be at least 0 to keep his reputation level. Rewritten it follows

$$\frac{R_{I,g}}{R_{I,b}} \geq \frac{1 - \gamma}{\gamma}.$$

Due to Equation (2.2) the reputation ratio is less than 1. This is only true for $\gamma \geq 0.5$ which is excluded by assumption (2.1). □

3 Proof of Theorem 3

Proof. To proof the condition $p \geq 1-\gamma$, I consider the contrary case $p < 1-\gamma$. In this case the change in reputation R is:

$$\begin{aligned} R &= p\gamma \cdot R_{I,g} - (1-p)(1-\gamma) \cdot R_{I,b} = \\ &= p \cdot (\gamma R_{I,g} + (1-\gamma) R_{I,b}) - (1-\gamma) R_{I,b} \\ &< (1-\gamma) \cdot (\gamma R_{I,g} + (1-\gamma) R_{I,b}) - (1-\gamma) \cdot R_{I,b} = \\ &= \gamma(1-\gamma) \cdot (R_{I,g} - R_{I,b}) \leq 0 \text{ for } 0 < \gamma < 0.5. \end{aligned}$$

In consequence, for the reputation change to be positive the minimum skill level for a certifier to play in the market is $p \geq 1-\gamma$.

A simpler proof is possible without considering reputation: A investor requires the PEF to certify at least as many good than bad PEFs (see Lemma 1). The number of certified good PEFs is $p\gamma$ and bad PEFs is $(1-p) \cdot (1-\gamma)$, so

$$\begin{aligned} p\gamma &\geq (1-p) \cdot (1-\gamma) \\ p\gamma - (1-p) \cdot (1-\gamma) &\geq 0 \\ p + \gamma - 1 &\geq 0 \\ p &\geq 1-\gamma. \end{aligned}$$

□

4 Proof of Lemma 4

Proof.

$$
\begin{aligned}
p\gamma \cdot R_{I,g} - (1-p)(1-\gamma) \cdot R_{I,b} &\geq 0 \\
\frac{R_{I,g}}{R_{I,b}} &\geq \frac{1-p}{p} \cdot \frac{1-\gamma}{\gamma} \\
\frac{p}{1-p} &\geq \frac{1-\gamma}{\gamma} \frac{R_{I,b}}{R_{I,g}} \text{ with } R_{I,g} > 0 \\
p &\geq (1-p) \cdot \frac{1-\gamma}{\gamma} \frac{R_{I,b}}{R_{I,g}} \\
p \cdot (1 + \frac{1-\gamma}{\gamma} \frac{R_{I,b}}{R_{I,g}}) &\geq \frac{1-\gamma}{\gamma} \frac{R_{I,b}}{R_{I,g}} \\
p &\geq \frac{\frac{1-\gamma}{\gamma} \frac{R_{I,b}}{R_{I,g}}}{1 + \frac{1-\gamma}{\gamma} \frac{R_{I,b}}{R_{I,g}}} \\
p &\geq \frac{1}{1 + \frac{\gamma}{1-\gamma} \cdot \frac{R_{I,g}}{R_{I,b}}}
\end{aligned}
$$

□

5 Proof of Theorem 5

Proof. The partial derivation of the PEFs' reputation change by p is always positive in the domain of definition:

$$
\begin{aligned}
\frac{\partial R_{PEF}}{\partial p} &= (\gamma - (1-\gamma)) \cdot R_{PEF,g} - (-\gamma + (1-\gamma)) \cdot R_{PEF,b} = \\
&= (2\gamma - 1) \cdot R_{PEF,g} + (2\gamma - 1) \cdot R_{PEF,b} \\
&\leq 0 \text{ for } \gamma < 0.5 \text{ and } R_{PEF,g} \geq 0 \text{ and } R_{PEF,b} \geq 0 \ .
\end{aligned}
$$

□

6 Proof that certification of more than 50% of private equity funds is not possible

Proof. This can be proven using contradiction. Let's assume

$$\begin{aligned} p\gamma + (1-p)(1-\gamma) &> p(1-\gamma) - (1-p)\gamma \\ 1 &> 2p + 2\gamma - 4p\gamma \\ 0.5 &> p + \gamma - 2p\gamma = p \cdot (1-2\gamma) + \gamma \end{aligned}$$

Using the condition $p \geq 1 - \gamma$ gives:

$$\begin{aligned} 0.5 &> (1-\gamma) \cdot (1-2\gamma) + \gamma \\ 0 &> \gamma^2 - \gamma + 0.25 \quad . \end{aligned}$$

This inequation is never correct for $\gamma < 0.5$, so certifying more than 50% of all PEFs is not possible. □

7 Proof that increasing p leads to a decrease in the share of certified private equity funds

Proof. Increasing p to $\hat{p} < 1$ reduces the number of certified PEFs. This can be proven by contradiction. I assume that $p < \hat{p}$. In this case the ratio in the number of certifiers is:

$$\begin{aligned} p\gamma + (1-p)(1-\gamma) &< \hat{p}\gamma + (1-\hat{p})(1-\gamma) \\ 2p\gamma - p &< 2\hat{p}\gamma - \hat{p} \\ p \cdot (2\gamma - 1) &< \hat{p} \cdot (2\gamma - \hat{p} \end{aligned}$$

With $\gamma < 0.5$ this results in $p > \hat{p}$ which is in contradiction with the assumption that $p < \hat{p}$. □

8 Proof that increasing p leads to an improvement of the ratio between good and bad certified private equity funds

Proof. Increasing p to $\hat{p} < 1$ increases the quality of certified PEFs. This can be proven by contradiction. I assume that the ratio between good and bad certified PEFs at p is higher than at $\hat{p}$:

$$\begin{aligned}
\frac{p\gamma}{(1-p)(1-\gamma)} &> \frac{\hat{p}\gamma}{(1-\hat{p})(1-\gamma)} \\
p\gamma(1-\hat{p})(1-\gamma) &> \hat{p}\gamma(1-p)(1-\gamma) \\
p(1-\hat{p}) &> \hat{p}(1-p) \\
p - p\cdot\hat{p} &> \hat{p} - p\cdot\hat{p}
\end{aligned}$$

With $0 < \gamma < 0.5$ this results in $p > \hat{p}$ which is in contradiction with the assumption that $\hat{p} > p$. So the quality of PEFs certified at $\hat{p}$ is higher than the quality at p. □

9 Proof that increasing γ to the supply curve level γ^* yields a higher compensation to a placement agent than decreasing p to p^* at γ_{min} for $p_{max}\gamma_{min} + (1-p_{max})(1-\gamma_{min}) < s \leq p_{max}\gamma^*(p_{max}) + (1-p_{max})(1-\gamma^*(p_{max}))$

Proof. The partial derivations of a placement agent's compensation function W_{PA} are

$$\begin{aligned}
\frac{\partial W_{PA}}{\partial p} &= (2\gamma - 1)\cdot W_F + \gamma \cdot W_C \\
\frac{\partial W_{PA}}{\partial \gamma} &= (2p - 1)\cdot W_F + p \cdot W_C > 0
\end{aligned}$$

for $s > p\gamma + (1-p)(1-\gamma)$, i. e. any p-γ-combination that does not or just clears the market. As $\frac{\partial W_{PA}}{\partial \gamma}$ shows, increasing γ is always lucrative for a PA. When $\frac{W_C}{W_F} > \frac{1-2\gamma}{\gamma}$ the partial derivation by p is positive and the PA maximizes p to p_{max} and γ to $\gamma^*(p_{max})$.

Question is if a placement agent prefers to maximize γ rather than minimize p for $\frac{W_C}{W_F} < \frac{1-2\gamma}{\gamma}$ (in this case the partial derivation $\frac{\partial W_{PA}}{\partial p}$ is negative). To answer this, I first assume that the PA minimizes p to the level on the supply curve at γ_{min}, i. e. $p^*(\gamma_{min})$. Starting from this point, the PA would have to accept an increase of p for any increase of γ to be able to stay on the supply curve. This is only an option for him if

$$\begin{aligned} \frac{\partial W_{PA}}{\partial p} &< \frac{\partial W_{PA}}{\partial \gamma} \\ (2\gamma - 1) \cdot W_F + \gamma \cdot W_C &< (2p-1) \cdot W_F + p \cdot W_C \end{aligned}$$

is true for any point on the supply curve. Replacing γ by the supply curve function $\gamma = \frac{s+p-1}{2p-1}$ and solving for W_F and W_C returns

$$\begin{aligned} (2 \cdot \frac{s+p-1}{2p-1} - 1) \cdot W_F + \frac{s+p-1}{2p-1} \cdot W_C < (2p-1) \cdot W_F + p \cdot W_C \\ 2 \cdot (1 + 2p^2 - 2p - s) \cdot W_F + (1 + 2p^2 - 2p - s) \cdot W_C > 0. \quad (1) \end{aligned}$$

Inequation (1) is true if s is less than $1+2p^2-2p$ in the domain of definition of p and s. $1 + 2p^2 - 2p$ takes values between 0.5 and 1 for $0.5 < p < 1$. Because s was defined to be less than 0.5 Inequation (1) is always true. So, a placement agent selects the highest possible γ that is in his option space on the supply curve for $\frac{W_C}{W_F} < \frac{1-2\gamma}{\gamma}$. □

Appendices for Chapter 3

Appendix

10 Sector Classification

Table 13: Overview on Sector Classification, Corresponding CEPRES Industry Classes and US Datastream Stock Price Indices

Sector classification	CEPRES classification	corresponding US Datastream price index
Biotechnology	Healthcare/LifeScience	Health Care
Industrials	Construction	Construction & Materials
	Consumer Industry/Food	Consumer Goods
	Environment	Utilities
	Hotel	Hotels
	Industrial/Manufacturing	Industrials
	Leisure	Leisure Goods
	Logistics	Industrial Transportation
	Materials	Basic Materials
	Natural Resources/Energy	Basic Resources
	Retail	Retail
	Textiles	Clothes & Accessoires
	Traditional Products	Industrials
	Transportation	Transportation Services
	Waste/Recycling	Waste & Displacement Services
Computer	High Tech	Technology
	IT	Computer Hardware
	Semiconductor	Semiconductors
Software	Software	Software
Internet & Media	Internet	Internet
	Media	Media
Telecommunication	Telecom	Telecom
Other & Unspecified	Financial Services	Financial Services
	Business Services	Business Support Services
	Other Services	Industrial Goods & Services
	Investment in preceding Fund	SP500
	Fund in Fund Investment	SP500
	Others	SP500
	Unspecified	SP500

Peter Lang · Internationaler Verlag der Wissenschaften

Bernhard Christian Kübler

Risk Classification by Means of Clustering

Frankfurt am Main, Berlin, Bern, Bruxelles, New York, Oxford, Wien, 2010.
IV, 195 pp., num tab. and graph.
Schriften zum Controlling, Finanz- und Risikomanagement.
Editor in charge: Andreas Brieden. Vol. 4
ISBN 978-3-631-59759-0 · hardback € 44,80*

Determining risk-adequate insurance premiums is a core issue in actuarial mathematics. This study is specifically concerned with identifying convenient partitions of (general) insurance collectives such that the resulting tariff classes are homogeneous to a maximum extent and – on the other hand – yet large enough to allow for the occurrence of the group balance concept and to end up with reliable estimates of the moments of the claim size distributions. Therefore, the author develops an innovative classification algorithm utilizing a multidimensional cluster approach combined with credibility-theoretical implications. Its construction stems from involving the entire claim information of risks simultaneously and in a suitable manner, and particulary from obtaining optimality regarding the cluster criterions. Under certain conditions, commonly used cross classification schemes are shown to be a particular case of the new approach. Besides desirable theoretical benefits like its generalizing established cross classification systems, an empirical investigation also suggests the practical superiority of the new algorithm.

Content: Application of Cluster Analysis in Actuarial Mathematics · Cross Classification Systems · Development of a New Classification System in order to Partition Insurance Collectives · Empirical Investigation: Comparison of the New Classification Algorithm and Cross Classification

Frankfurt am Main · Berlin · Bern · Bruxelles · New York · Oxford · Wien
Distribution: Verlag Peter Lang AG
Moosstr. 1, CH-2542 Pieterlen
Telefax 00 41 (0) 32 / 376 17 27